# Lecture Notes in Statistics

Edited by D. Brillinger, S. Fienberg, J. Gani, J. Hartigan, J. Kiefer, and K. Krickeberg

BT!

[...SYMP. ON MATH. STATS.]

8



Pál Révész
Leopold Schmetterer
V. M. Zolotarev



# The First Pannonian Symposium on Mathematical Statistics

Springer-Verlag
New York Heidelberg Berlin

Pál Révész
Mathematical Institute of
the Hungarian Academy
of Sciences
1053 Budapest V.
Realtanoda 13-15
Hungary

Leopold Schmetterer
Inst. F. Statistik
Universität Wien
Rathausstr. 19
A-1010 Wien
Austria

V. M. Zolotarev
Steklov, Mathematical
Institute
ul Vavilova 42
Moscow, 117333
USSR



AMS Classification: 62-06

**Library of Congress Cataloging in Publication Data**

Pannonian Symposium on Mathematical Statistics
(1st: 1979: Bad Tatzmannsdorf, Austria)
The First Pannonian Symposium on Mathematical
Statistics.

(Lecture notes in statistics; 8)
1. Mathematical statistics—Congresses.
I. Révész, Pál. II. Schmetterer, Leopold, 1919–
III. Zolotarev, V. M. IV. Title. V. Series:
Lecture notes in statistics (Springer-Verlag); v. 8.
QA276.A1P36 1979 519.5 81–4060
AACR2



Printed in the United States of America

9 8 7 6 5 4 3 2 1

ISBN 0-387-90583-9 Springer-Verlag New York Heidelberg Berlin
ISBN 3-540-90583-9 Springer-Verlag Berlin Heidelberg New York

# PREFACE

The first Pannonian Symposium on Mathematical Statistics was held at Bad Tatzmannsdorf (Burgenland/Austria) from September 16th to 21st, 1979. The aim of it was to further and intensify scientific cooperation in the Pannonian area, which, in a broad sense, can be understood to cover Hungary, the eastern part of Austria, Czechoslovakia, and parts of Poland, Yugoslavia and Romania. The location of centers of research in mathematical statistics and probability theory in this territory has been a good reason for the geographical limitation of this meeting.

About 70 researchers attended this symposium, and 49 lectures were delivered; a considerable part of the presented papers is collected in this volume. Beside the lectures, vigorous informal discussions among the participants took place, so that many problems were raised and possible ways of solutions were attacked.

We take the opportunity to thank Dr. U. Dieter (Graz), Dr. F. Konecny (Wien), Dr. W. Krieger (Göttingen) and Dr. E. Neuwirth (Wien) for their valuable help in the refereeing work for this volume. The Pannonian Symposium could not have taken place without the support of several institutions: The Austrian Ministry for Research and Science, the State government of Burgenland, the Community Bad Tatzmannsdorf, the Kurbad Tatzmannsdorf AG, the Austrian Society for Information Science and Statistics, IBM Austria, Volksbank Oberwart, Erste Osterreichische Spar-Casse and Spielbanken AG Austria. The Austrian Academy of Sciences made possible the participation in the Symposium for several mathematicians. We express our gratitude to all these institutions for their generous help.

On behalf of the Organizing Committee

Wilfried Grossmann Georg Pflug Wolfgang Wertz

## CONTENTS

# IMPROVEMENT OF EXTRAPOLATION IN MULTIPLE TIME SERIES

Jiří Anděl

Charles University, Prague

## 1. Introduction

Let $\{X_t\}$ be a p-dimensional random process with zero expectation and with finite second moments. We shall investigate the discrete case, when $t=\ldots,-1,0,1,\ldots$ . Assume that the vectors $X_{t-1},X_{t-2},\ldots$ are known and that we wish to get the best linear extrapolation $\hat{X}_t$ of the vector $X_t$. Our extrapolation $\hat{X}_t$ can be calculated either by a method given in Rozanov [3] or by a well known iterative procedure, which is briefly described in our Section 2. The accuracy of $\hat{X}_t$ is usually measured by the residual variance matrix

$$\Delta_X = E(X_t-\hat{X}_t)(X_t-\hat{X}_t)',$$

where the prime denotes the transposition. If the diagonal elements of the matrix $\Delta_X$ are too large, the extrapolation $\hat{X}_t$ is not satisfactory and we should like to improve it. It may happen that we can observe another discrete q-dimensional process $\{Y_t\}$, also with zero expectation and with finite second moments. If $\{Y_t\}$ is correlated with $\{X_t\}$, we can use the information contained in $\{Y_t\}$ and improve the extrapolation of the process $\{X_t\}$. Denote $W_t=(X_t',Y_t')'$. The simplest case is that we know vectors $W_{t-1},W_{t-2},\ldots$, so that it is possible to calculate the best linear extrapolation $\hat{W}_t$ of the vector $W_t$ in an usual way. The first p components of $\hat{W}_t$ will be denoted by $\hat{X}_t(1,1)$. We can also say that $\hat{X}_t(1,1)$ is the best linear extrapolation of $X_t$ in the case that the vectors $X_{t-1},Y_{t-1},X_{t-2},Y_{t-2},\ldots$ are known. Surprisingly, it may happen even in "regular" models that $\hat{X}_t= \hat{X}_t(1,1)$; see Anděl [1].

We shall restrict ourselves to the case when $\{W_t\}$ is an ARMA process with known parameters. Consider the situation that an extrapolation of $X_t$ must be calculated when not only $X_{t-1}, X_{t-2}, \ldots$ and $Y_{t-1}, Y_{t-2}, \ldots$ are known, but also $Y_t$ (or $Y_t$ and $Y_{t+1}$) are given. We shall derive the corresponding formulas for extrapolation and we shall also find the conditions under which $Y_t$ (or $Y_{t+1}$) cannot improve the extrapolation of the vector $X_t$. Our results generalize the formulas in Anděl [2] , where the case p=q=1 was investigated.

2. Preliminaries

The methods for solving our problems are based on theory of Hilbert space and on matrix theory. In the first part of this Section we introduce some auxiliary assertions from this area.

Theorem 1. Let y, $x_t$ and $z_s$ (where t and s are some integers) be elements of a Hilbert space H. Denote $H\{u,v,\ldots\}$ the Hilbert subspace spanned by elements $u,v,\ldots \in H$. If $\hat{y}$ is the projection of y on $H\{\{x_t\},\{z_s\}\}$, $\tilde{y}$ is the projection of y on $H\{\{x_t\}\}$, $\tilde{z}_s$ is the projection of $z_s$ on $H\{\{x_t\}\}$ and $\bar{y}$ is the projection of y on $H\{\{z_s-\tilde{z}_s\}\}$, then

$$\hat{y} = \tilde{y} + \bar{y}.$$

Proof is left out, since the assertion is well known.

In statistical applications, the elements of a Hilbert space are usually random variables. Their projections are calculated by means of regression methods.

Theorem 2. Let X and Y be two random vectors with finite second moments. Let H be the Hilbert space spanned by their elements, where the scalar product is defined by the covariance. Let $H_1$ be the subspace of H, spanned by the elements of Y. If Var Y is a regular matrix, then the vector $\hat{X}$ of projections of elements of X on $H_1$ is given by formula

$$\hat{X} = \mathrm{Cov}(X,Y)(\mathrm{Var}\ Y)^{-1}Y.$$

Proof. Theorem 2 is known from linear regression theory. Let us only emphasize that the dimensions of X and Y need not be the same.

Theorem 3. Let H be a Hilbert space and $H_1 \supset H_2$ two its subspaces. For $y \in H$ denote $y_i$ the projection of y on $H_i$, i=1,2. If $y_1 \neq y_2$, then

$$\| y-y_1 \| < \| y-y_2 \| .$$

Proof of this elementary assertion can be found in Anděl [2].

Theorem 3 justifies the following consideration. Let us have two sets $H_1$ and $H_2$ of random variables, $H_1 \supset H_2$. Let X be a random variable and $X_1$, $X_2$ its best linear estimates based on $H_1$, $H_2$, respectively. If $X_1 \neq X_2$, then $X_1$ is strongly better than $X_2$ in the sense that $\mathrm{Var}(X-X_1) < \mathrm{Var}(X-X_2)$.

Now, let us define a Hilbert space spanned by random vectors $\{U_t\}_{t \in T}$. If the vectors $U_t$ have finite second moments, then $H\{U_t\}_{t \in T}$ denotes the Hilbert space spanned by all elements of vectors $U_t$ for all $t \in T$. If U is a random vector, then the symbol $U \in H\{U_t\}_{t \in T}$ means that every element of U belongs to $H\{U_t\}_{t \in T}$.

Theorem 4. Let $S = \begin{Vmatrix} K, & L \\ M, & N \end{Vmatrix}$ be a matrix written in blocks, where N is a regular $q \times q$ block. Then the rank of S is q if and only if

$$K-LN^{-1}M=0.$$

Proof. Denote I the unit matrix. The matrices

$$G_1 = \begin{Vmatrix} I, & -LN^{-1} \\ 0, & I \end{Vmatrix}, \qquad G_2 = \begin{Vmatrix} I, & 0 \\ -N^{-1}M, & I \end{Vmatrix}$$

are regular and, therefore, the rank of the matrix

$$G_1 S G_2 = \begin{Vmatrix} K-LN^{-1}M, & 0 \\ 0, & N \end{Vmatrix}$$

is the same as the rank of S. Since N has rank q, $G_1SG_2$ has also rank q if and only if $K-LN^{-1}M=0$.

## 3. Improvement of extrapolation in multiple ARMA models

Let $\{Z_t\}$ be (p+q)-dimensional uncorrelated random vectors with $EZ_t=0$, Var $Z_t=I$. We shall write $Z_t=(Z_t^{1\prime}, Z_t^{2\prime})'$, where $Z_t^1$ has p components and $Z_t^2$ has q components. Let $\{W_t\}$ be a (p+q)-dimensional

ARMA process defined by

$$(1) \qquad \sum_{k=0}^{m} A_k W_{t-k} = \sum_{j=0}^{n} B_j Z_{t-j} ,$$

where $A_k$ and $B_j$ are $(p+q) \times (p+q)$ matrices such that

$$(2) \qquad \det\Big(\sum_{k=0}^{m} A_k z^k\Big) \neq 0 \quad \text{for} \quad |z| \leqq 1$$

and

$$(3) \qquad \det\Big(\sum_{j=0}^{n} B_j z^j\Big) \neq 0 \quad \text{for} \quad |z| \leqq 1.$$

It follows from (2) and (3) that matrices $A_0$ and $B_0$ must be regular. Condition (2) ensures the existence of a process $\{W_t\}$ in (1). An ARMA process $\{W_t\}$ satisfying (3) is called invertible. It is well known that if (2) holds, then

$$(4) \qquad W_t = \sum_{k=0}^{\infty} C_k Z_{t-k},$$

where $C_k$ are $(p+q) \times (p+q)$ matrices. Similarly, condition (3) enables to derive that

$$(5) \qquad Z_t = \sum_{k=0}^{\infty} D_k W_{t-k},$$

where $D_k$ are also $(p+q) \times (p+q)$ matrices. The series in (4) as well as in (5) converge in the quadratic mean for every component. Obviously, we have from (4) and (5) that

$$H\{X_t\}_{t \leqq s} = H\{Z_t\}_{t \leqq s}$$

holds for every integer s.

Without any loss of generality we shall assume that $A_0 = I$. Denote $\hat{W}_t$ the best linear extrapolation of $W_t$ based on $W_{t-1}, W_{t-2}, \dots$ Our assumptions imply that

$$(6) \qquad W_t = B_0 Z_t + \hat{W}_t,$$

where

$$(7) \qquad \hat{W}_t = -\sum_{k=1}^{m} A_k W_{t-k} + \sum_{j=1}^{n} B_j Z_{t-j}.$$

For practical purposes we rewrite (6) into the form

$$(8) \qquad Z_t = B_0^{-1}(W_t - \hat{W}_t),$$

which holds for every t. Inserting into (7) we obtain

$$(9) \qquad \hat{W}_t = -\sum_{k=1}^{m} A_k W_{t-k} + \sum_{j=0}^{n} B_j B_0^{-1}(W_{t-j} - \hat{W}_{t-j})$$

and this is a variant of the recurrent formula, which is usually used for extrapolation in ARMA models. A similar formula can be easily derived also for extrapolation of $W_{t+u}$ when $u \geq 1$.

Denote $W_t = (X_t', Y_t')'$, where $X_t$ and $Y_t$ have p and q components, respectively. Let $\hat{X}_t(a,b)$ be the best linear extrapolation of the vector $X_t$ based on $X_{t-a}, X_{t-a-1}, \ldots, Y_{t-b}, Y_{t-b-1}, \ldots$ . Put

$$\Delta_X(a,b) = E[X_t - \hat{X}_t(a,b)][X_t - \hat{X}_t(a,b)]' .$$

Obviously,

$$\hat{W}_t = \begin{Vmatrix} \hat{X}_t(1,1) \\ \hat{Y}_t(1,1) \end{Vmatrix} .$$

We shall assume that $\hat{X}_t(1,1)$ and $\hat{Y}_t(1,1)$ are known, since they can be easily computed from (9). Let us consider the problem how to improve $\hat{X}_t(1,1)$, if the vector $Y_t$ is also known (or if $Y_t$ and $Y_{t+1}$ are known). It means that we want to calculate $\hat{X}_t(1,0)$ $[$or $\hat{X}_t(1,-1)]$. We shall also investigate under which conditions the equalities

$$\hat{X}_t(1,1) = \hat{X}_t(1,0) \qquad \text{and} \qquad \hat{X}_t(1,0) = \hat{X}_t(1,-1)$$

hold. To simplify the notation in the next formulas, we put $A_k = 0$ for $k \notin \{0,1,\ldots,m\}$ and $B_j = 0$ for $j \notin \{0,1,\ldots,n\}$. For every k and j write

$$(10) \qquad A_k = \begin{Vmatrix} A_k^{11}, & A_k^{12} \\ A_k^{21}, & A_k^{22} \end{Vmatrix}, \qquad B_j = \begin{Vmatrix} B_j^{11}, & B_j^{12} \\ B_j^{21}, & B_j^{22} \end{Vmatrix},$$

where $A_k^{11}$ and $B_j^{11}$ are $p\times p$ blocks, whereas $A_k^{22}$ and $B_j^{22}$ are $q\times q$ blocks.

Theorem 5. If $B_0^{22}$ is a regular matrix, then

$$\hat{X}_t(1,0) = \hat{X}_t(1,1) + \bar{y}, \tag{11}$$

where

$$\bar{y} = (B_0^{11}B_0^{21\prime}+B_0^{12}B_0^{22\prime})(B_0^{21}B_0^{21\prime}+B_0^{22}B_0^{22\prime})^{-1}[Y_t-\hat{Y}_t(1,1)], \tag{12}$$

and the equality $\Delta_X(1,1)=\Delta_X(1,0)$ holds if and only if

$$B_0^{11}B_0^{21\prime}+B_0^{12}B_0^{22\prime} = 0. \tag{13}$$

Proof. According to Theorems 1 and 2 we obtain that (11) holds, with

$$\bar{y} = \mathrm{Cov}[X_t,Y_t-\hat{Y}_t(1,1)]\,\{\mathrm{Var}[Y_t-\hat{Y}_t(1,1)]\}^{-1}[Y_t-\hat{Y}_t(1,1)]. \tag{14}$$

From (6) we get

$$X_t = B_0^{11}Z_t^1 + B_0^{12}Z_t^2 + \hat{X}_t(1,1), \tag{15}$$

$$Y_t - \hat{Y}_t(1,1) = B_0^{21}Z_t^1 + B_0^{22}Z_t^2. \tag{16}$$

Clearly,

$$\mathrm{Var}[Y_t-\hat{Y}_t(1,1)] = B_0^{21}B_0^{21\prime} + B_0^{22}B_0^{22\prime}. \tag{17}$$

Since $\mathrm{Cov}[\hat{X}_t(1,1),Y_t-\hat{Y}_t(1,1)] = 0$, formula (12) easily follows from (14), (15) and (16). The rest of Theorem 5 is a consequence of Theorem 3.

From practical point of view, it is not necessary to insert for $Y_t-\hat{Y}_t(1,1)$ into (12) from (16), because we have assumed that $Y_t$ is known and that $\hat{Y}_t(1,1)$ had been calculated before.

Theorem 6. Let $B_0^{22}$ be a regular matrix. Denote

$$F_1=B_1^{21}-A_1^{21}B_0^{11}-A_1^{22}B_0^{21}, \qquad F_2=B_1^{22}-A_1^{21}B_0^{12}-A_1^{22}B_0^{22}, \tag{18}$$

$$G=B_0^{21}B_0^{21\prime}+B_0^{22}B_0^{22\prime}. \tag{19}$$

Then

$$\hat{X}_t(1,-1) = \hat{X}_t(1,0) + \bar{y}, \tag{20}$$

where

(21) $\bar{y} = \left[B_0^{11}F_1'+B_0^{12}F_2'-(B_0^{11}B_0^{21\prime}+B_0^{12}B_0^{22\prime})G^{-1}(B_0^{21}F_1'+B_0^{22}F_2')\right]\cdot$

$\left[G+F_1F_1'+F_2F_2'-(F_1B_0^{21\prime}+F_2B_0^{22\prime})G^{-1}(B_0^{21}F_1'+B_0^{22}F_2')\right]^{-1}\cdot$

$\left\{Y_{t+1}-\hat{Y}_{t+1}(2,2)-(F_1B_0^{21\prime}+F_2B_0^{22\prime})G^{-1}[Y_t-\hat{Y}_t(1,1)]\right\}$

and the equality $\Delta_X(1,0) = \Delta_X(1,-1)$ holds if and only if

(22) $$F_1-F_2(B_0^{22})^{-1}B_0^{21} = 0.$$

Proof. Applying Theorems 1 and 2 we get that (20) holds where

(23) $\bar{y}=\mathrm{Cov}[X_t,Y_{t+1}-\hat{Y}_{t+1}(2,1)]\{\mathrm{Var}[Y_{t+1}-\hat{Y}_{t+1}(2,1)]\}^{-1}[Y_{t+1}-\hat{Y}_{t+1}(2,1)].$

Analogously, we obtain

(24) $$\hat{Y}_{t+1}(2,1) = \hat{Y}_{t+1}(2,2) + y^*,$$

where

(25) $y^*= \mathrm{Cov}[Y_{t+1},Y_t-\hat{Y}_t(1,1)]\{\mathrm{Var}[Y_t-\hat{Y}_t(1,1)]\}^{-1}[Y_t-\hat{Y}_t(1,1)].$

If we use formula (6), we come to

(26) $$W_{t+1} = B_0Z_{t+1} + (B_1-A_1B_0)Z_t + \tilde{W}_{t+1},$$

where $\tilde{W}_{t+1}= \sum_{j=3}^{n} B_jZ_{t+1-j} - A_1\hat{W}_t$ is the best linear extrapolation of $W_{t+1}$ based on $W_{t-1},W_{t-2},\ldots$ Especially, formula (26) gives

(27) $$Y_{t+1} = B_0^{21}Z_{t+1}^1+B_0^{22}Z_{t+1}^2+F_1Z_t^1+F_2Z_t^2+\hat{Y}_{t+1}(2,2).$$

From (16), (25) and (27) we have

(28) $$y^* = (F_1B_0^{21\prime}+F_2B_0^{22\prime})G^{-1}[Y_t-\hat{Y}_t(1,1)] .$$

According to (24),

(29) $$Y_{t+1}-\hat{Y}_{t+1}(2,1) = Y_{t+1}-\hat{Y}_{t+1}(2,2)-y^* ,$$

and using (16), (27) and (28) we can write

$$(30) \qquad Y_{t+1}-\hat{Y}_{t+1}(2,1) = B_0^{21}Z_{t+1}^1+B_0^{22}Z_{t+1}^2+$$
$$\left[F_1-(F_1B_0^{21\prime}+F_2B_0^{22\prime})G^{-1}B_0^{21}\right]Z_t^1+$$
$$\left[F_2-(F_1B_0^{21\prime}+F_2B_0^{22\prime})G^{-1}B_0^{22}\right]Z_t^2.$$

From (15) and (30) we get

$$(31) \quad \mathrm{Cov}\left[X_t,Y_{t+1}-\hat{Y}_{t+1}(2,1)\right] = B_0^{11}\left[F_1-(F_1B_0^{21\prime}+F_2B_0^{22\prime})G^{-1}B_0^{21}\right]'+$$
$$B_0^{12}\left[F_2-(F_1B_0^{21\prime}+F_2B_0^{22\prime})G^{-1}B_0^{22}\right]'=$$
$$B_0^{11}F_1'+B_0^{12}F_2'-(B_0^{11}B_0^{21\prime}+B_0^{12}B_0^{22\prime})G^{-1}(B_0^{21}F_1'+B_0^{22}F_2'),$$

whereas (30) itself leads after some computations to

$$(32) \quad \mathrm{Var}\left[Y_{t+1}-\hat{Y}_{t+1}(2,1)\right] = G+F_1F_1'+F_2F_2'-(F_1B_0^{21\prime}+F_2B_0^{22\prime})G^{-1}(B_0^{21}F_1'+B_0^{22}F_2').$$

Now, (21) follows from (23), (28), (29), (31) and (32).

The equality $\Delta_X(1,0)=\Delta_X(1,-1)$ holds if and only if $\mathrm{Cov}\left[X_t,Y_{t+1}-\hat{Y}_{t+1}(2,1)\right] = 0$. Consider matrix

$$S = \begin{Vmatrix} F_1B_0^{11\prime}+F_2B_0^{12\prime}, & F_1B_0^{21\prime}+F_2B_0^{22\prime} \\ B_0^{21}B_0^{11\prime}+B_0^{22}B_0^{12\prime}, & B_0^{21}B_0^{21\prime}+B_0^{22}B_0^{22\prime} \end{Vmatrix}.$$

We have

$$(33) \qquad S = FB_0',$$

where

$$(34) \qquad F = \begin{Vmatrix} F_1, & F_2 \\ B_0^{21}, & B_0^{22} \end{Vmatrix} .$$

We shall apply Theorem 4. Expression (31) (or, more precisely, its transposition) is the zero matrix if and only if the rank of S is q. Since $B_0$ is regular, the rank of S is the same as the rank of F. But the rank of F is q if and only if (22) holds.

For some purposes, it is also interesting to have a formula for

$\hat{X}_t(1,-2)$. The final expression would be very cumbersome and, therefore, we shall describe the procedure in a few steps.

Theorem 7. The expression $\hat{X}_t(1,-2)$ is given by

$$\hat{X}_t(1,-2) = \hat{X}_t(1,-1) + \bar{y}, \tag{35}$$

where

$$\bar{y}=\mathrm{Cov}\left[X_t, Y_{t+2}-\hat{Y}_{t+2}(3,1)\right]\left\{\mathrm{Var}\left[Y_{t+2}-\hat{Y}_{t+2}(3,1)\right]\right\}^{-1}\left[Y_{t+2}-\hat{Y}_{t+2}(3,1)\right]. \tag{36}$$

For explicit calculations we can use the formula

$$Y_{t+2}-\hat{Y}_{t+2}(3,1)=Y_{t+2}-\hat{Y}_{t+2}(3,3)-y^*-y^o, \tag{37}$$

where

$$y^o=\mathrm{Cov}\left[Y_{t+2}, Y_{t+1}-\hat{Y}_{t+1}(2,1)\right]\left\{\mathrm{Var}\left[Y_{t+1}-\hat{Y}_{t+1}(2,1)\right]\right\}^{-1}\left[Y_{t+1}-\hat{Y}_{t+1}(2,1)\right] \tag{38}$$

and

$$y^*=\mathrm{Cov}\left[Y_{t+2}, Y_t-\hat{Y}_t(1,1)\right]\left\{\mathrm{Var}\left[Y_t-\hat{Y}_t(1,1)\right]\right\}^{-1}\left[Y_t-\hat{Y}_t(1,1)\right]. \tag{39}$$

Proof. Theorems 1 and 2 give immediately

$$\hat{X}_t(1,-2)=\hat{X}_t(1,-1)+\bar{y},\quad \hat{Y}_{t+2}(3,1)=\hat{Y}_{t+2}(3,2)+y^o,\quad \hat{Y}_{t+2}(3,2)=\hat{Y}_{t+2}(3,3)+y^*,$$

where $\bar{y}$, $y^o$ and $y^*$ are introduced in (36), (38) and (39), respectively. From here the assertion is obvious.

We shall show how to calculate the expressions in Theorem 7 which were not given before. Using three times formula (6), we have

$$W_{t+2}=B_0Z_{t+2}+(B_1-A_1B_0)Z_{t+1}+\left[B_2-A_1B_1+(A_1^2-A_2)B_0\right]Z_t+W^*_{t+2}, \tag{40}$$

where $W^*_{t+2}$ is the best linear extrapolation of $W_{t+2}$ based on $W_{t-1}$, $W_{t-2}, \ldots$ Denote

$$H_1=B_2^{21}-A_1^{21}B_1^{11}-A_1^{22}B_1^{21}, \qquad H_2=B_2^{22}-A_1^{21}B_1^{12}-A_1^{22}B_1^{22}, \tag{41}$$

$$K_1=A_2^{21}-A_1^{21}A_1^{11}-A_1^{22}A_1^{21}, \qquad K_2=A_2^{22}-A_1^{21}A_1^{12}-A_1^{22}A_1^{22}. \tag{42}$$

Then

$$(43)\quad Y_{t+2}=\hat{Y}_{t+2}(3,3)+B_0^{21}Z_{t+2}^1+B_0^{22}Z_{t+2}^2+F_1Z_{t+1}^1+F_2Z_{t+1}^2+$$

$$(H_1-K_1B_0^{11}-K_2B_0^{21})Z_t^1+(H_2-K_1B_0^{12}-K_2B_0^{22})Z_t^2,$$

where $F_1$ and $F_2$ are given in (18). Using (16), we get

$$(44)\quad \mathrm{Cov}\left[Y_{t+2},Y_t-\hat{Y}_t(1,1)\right]=(H_1-K_1B_0^{11}-K_2B_0^{21})B_0^{21\prime}+(H_2-K_1B_0^{12}-K_2B_0^{22})B_0^{22\prime}.$$

Analogously, (30) gives

$$(45)\quad \mathrm{Cov}\left[Y_{t+2},Y_{t+1}-\hat{Y}_{t+1}(2,1)\right]=F_1B_0^{21\prime}+F_2B_0^{22\prime}+$$

$$(H_1-K_1B_0^{11}-K_2B_0^{21})\left[F_1-(F_1B_0^{21\prime}+F_2B_0^{22\prime})G^{-1}B_0^{21}\right]'+$$

$$(H_2-K_1B_0^{12}-K_2B_0^{22})\left[F_2-(F_1B_0^{21\prime}+F_2B_0^{22\prime})G^{-1}B_0^{22}\right]'.$$

Formulas (44) and (45) together with (17) and (32) enable to calculate $y^*$ and $y^o$. To derive $\bar{y}$, it is necessary to insert from (43) into (37), which is quite easy if the matrices $A_k$ and $B_j$ are given numerically, but not very elegant for a general model. Some explicit results can be reached for p=q=1 (see Anděl [2] ).

4. Special cases

Theorem 8. Let $\{W_t\}$ be a stationary AR(1) process defined by

$$W_t+AW_{t-1}=BZ_t.$$

Put

$$A=\left\|\begin{matrix}A^{11}, & A^{12}\\ A^{21}, & A^{22}\end{matrix}\right\|,\qquad B=\left\|\begin{matrix}B^{11}, & B^{12}\\ B^{21}, & B^{22}\end{matrix}\right\|,$$

where $A^{11}$ and $B^{11}$ are $p\times p$ blocks, and $A^{22}$ and $B^{22}$ are $q\times q$ blocks. If $B^{22}$ is a regular matrix, then the following assertions hold:

(i) $\left[\Delta_X(1,1)=\Delta_X(1,0)\right] \iff \left[B^{11}B^{21\prime}+B^{12}B^{22\prime}=0\right]$;

(ii) $\left[\Delta_X(1,0)=\Delta_X(1,-1)\right] \iff \left[A^{21}=0\right]$;

(iii) $\left[\Delta_X(1,0)=\Delta_X(1,-1)\right] \Rightarrow \left[\Delta_X(1,-1)=\Delta_X(1,-2)\right]$.

**Proof.** (i) This assertion was proved in Theorem 5.

(ii) According to (22), the condition $\Delta_X(1,0)=\Delta_X(1,-1)$ holds if and only if

$$-A^{21}B^{11}-A^{22}B^{21}+(A^{21}B^{12}+A^{22}B^{22})(B^{22})^{-1}B^{21} = 0,$$

which is the same as

$$-A^{21}\left[B^{11}-B^{12}(B^{22})^{-1}B^{21}\right] = 0.$$

Since B as well as $B^{22}$ are assumed to be regular, $B^{11}-B^{12}(B^{22})^{-1}B^{21}$ must be also regular and we have $A^{21}=0$.

(iii) We use the notation introduced in Theorem 7 and in its proof. The assumption $\Delta_X(1,0)=\Delta_X(1,-1)$ is equivalent with $A^{21}=0$ as well as with $\mathrm{Cov}\left[X_t,Y_{t+1}-\hat{Y}_{t+1}(2,1)\right]=0$ in view of (20) and (23). Because of (35), (36) and (37), we see that in this case the equality $\Delta_X(1,-1)=\Delta_X(1,-2)$ holds if

$$\mathrm{Cov}\left[X_t,Y_{t+2}-\hat{Y}_{t+2}(3,3)\right] - \mathrm{Cov}(X_t,y^*) = 0. \tag{46}$$

Since in our case $H_1=0$, $H_2=0$, $K_1=0$, $K_2=-A^{22}A^{22}$, we obtain

$$\mathrm{Cov}\left[X_t,Y_{t+2}-\hat{Y}_{t+2}(3,3)\right] = (B^{11}B^{21\prime}+B^{12}B^{22\prime})(A^{22}A^{22})'$$

and (after some elementary computations)

$$\mathrm{Cov}(X_t,y^*) = (B^{11}B^{21\prime}+B^{12}B^{22\prime})(A^{22}A^{22})',$$

and so (46) clearly holds.

The result (iii) can be interpreted in this way: If in AR(1) model the vector $Y_{t+1}$ cannot improve the extrapolation $X_t(1,0)$, then $Y_{t+2}$ cannot improve it either.

<u>Theorem 9</u>. Let $\{W_t\}$ be an invertible MA(1) process defined by $W_t=B_0Z_t+B_1Z_{t-1}$, where $B_0^{22}$ is a regular matrix. Then

(i) $\left[\Delta_X(1,1) = \Delta_X(1,0)\right] \Leftrightarrow \left[B_0^{11}B_0^{21\prime}+B_0^{12}B_0^{22\prime} = 0\right]$;

(ii) $\left[\Delta_X(1,0)=\Delta_X(1,-1)\right] \Leftrightarrow \left[B_1^{21}-B_1^{22}(B_0^{22})^{-1}B_0^{21}=0\right]$;

(iii) $\left[\Delta_X(1,0)=\Delta_X(1,-1)\right] \Rightarrow \left[\Delta_X(1,-1)=\Delta_X(1,-2)\right]$.

Proof. (i) See Theorem 5.

(ii) This assertion follows from (22), because in our case $F_1=B_1^{21}$, $F_2=B_1^{22}$.

(iii) From $\Delta_X(1,0)=\Delta_X(1,-1)$ we have $\operatorname{Cov}\left[X_t, Y_{t+1}-\hat{Y}_{t+1}(2,1)\right]=0$. Since $\hat{Y}_{t+2}(3,3)=0$, $y^*=0$ and $\operatorname{Cov}(X_t,Y_{t+2})=0$, we obtain from (36) and (37)

$$\operatorname{Cov}\left[X_t, Y_{t+2}-\hat{Y}_{t+2}(3,1)\right]=0,$$

i.e. $\bar{y}=0$ and, therefore, $\Delta_X(1,-2)=\Delta_X(1,-1)$.

The interpretation of (iii) is the same as that of (iii) in Theorem 8.

## References

[1] Anděl J. (1979). Measures of dependence in discrete stationary processes. Math. Operationsforsch. Statist., Ser. Statistics 10, 107-126.

[2] Anděl J. (1979). On extrapolation in two-dimensional stationary processes. To appear.

[3] Rozanov Ju. A. (1963). Stacionarnyje slučajnyje processy. Gos. izd., Moskva.

# ALGORITHMICAL DEFINITION OF FINITE BINARY RANDOM SEQUENCE

Dragan Banjević and Zoran Ivković

University of Belgrade

In [5] A.N. Kolmogorov has defined the complexity $K_F(x)$ of binary sequence $x$ with respect to program $F$ for the description of $x$. He prooved the existance of a universal program $F_0$ i.e. of such a program that $K_{F_0}(x) \leq K_F(x) + C_F$, $C_F$ -const., for each $x$ and $F$. Loosely speaking, the sequence $x$ is random if $K_{F_0}(x)$ is large enough. It is clear that, if the set $\mathcal{T}$ of sequences $x$ is finite, then each program $F$ is universal. This means that, in this case, the randomness can be defined arbitrary. For that reason this approach is more successful in case of infinite set $\mathcal{T}$ (see, for instance [4]).

However, Kolmogorov himself, in one of his previous papers [3], gave the definition of finite random sequence. This definition is based on the ideas of von Mises, improoved by Church [2]. It seems that this approach is interesting for two reasons: it is in accordance with the statistical interpretation of probability, and with algorithmical testing, which is important for applications (tables of random numbers). Except in the case [3], we do not know any other papers approaching this question.

The idea of this randomness of a binary sequence $T = (t_1, \ldots, t_N)$, $t_i \in \{0,1\}$, is in considering the rule $R$ for selection of the subsequence $S = (t_{i_1}, \ldots, t_{i_\nu})$ and in considering the relative frequency $f_1 = \frac{1}{\nu} \sum_{j=1}^{\nu} t_{i_j}$ of the accurancy of the number one in $S$. The sequence $T$ is $p$-random ($p$-fihed number, $0 \leq p \leq 1$), if $f_1$ is "close" to $p$. The rule $R$ is given by a sequence of computable functions $L_1(T), \ldots, L_N(T)$. Member $t_i$ is selected for $S$ if $L_i(T) = 1$.

The idea of von Mises, or better of Kolmogorov, is in

following: a) the decision of selection $t_i$ for $S$ does not depend on $t_i$ ,i.e. $L_i(t_1,\dots,t_i,\dots,t_N)=L_i(t_1,\dots,1-t_i,\dots,t_N)$, b) the order of decisions is sequential, in the sense that the decision of selection $t_i$ for $S$ depends only on previously examined members of $T$ .The following Kolmogorov's definition of the rule or algorithm $R$ for selection, completely reflects this idea: the algorithm $R$ is determined by a system of computable functions $F=(F_0,\dots,F_{N-1})$, $G=(G_0,\dots,G_{N-1})$, $H=(H_0,\dots,H_N)$; $F_i\in\{1,\dots,N\}$; $G_i,H_i\in\{0,1\}$, $F_0,G_0,H_0$-const., $H_N\equiv 1$.

The functions $F$ determine the examination order of members of $T$ , i.e. they define a permutation $(x_1,\dots,x_N)$ of $(1,\dots,N)$ by $x_i=F_{i-1}(x_1,t_{x_1};\dots;x_{i-1},t_{x_{i-1}})$ , $i=1,\dots,N$ . The functions $H$ determine the stopping rule: the last examined member is $t_{x_s}$ if $H_i(x_1,t_{x_1};\dots;x_i,t_{x_i})=0$ for $i<s$ and $H_s(x_1,t_{x_1};\dots;x_s,t_{x_s})=1$ .The functions $G$ determine the selection of member of $T$ for $S$ : $t_{x_i}$ is selected for $S$ if $G_{i-1}(x_1,t_{x_1};\dots;x_{i-1},t_{x_{i-1}})=1$ and $i\leq s$ .

It is easy to see that, instead of functions $G_{i-1}$ from $G$ we can consider $\tilde{G}_{i-1}(x_1,t_{x_1};\dots;x_{i-1},t_{x_{i-1}};x_i)$ because of

$G_{i-1}(x_1,t_{x_1};\dots;x_{i-1},t_{x_{i-1}})=\tilde{G}_{i-1}(x_1,t_{x_1};\dots;x_{i-1},t_{x_{i-1}},F_{i-1})$. On the other side, without loss of generality, we can substitute the system $(F,G,H)$ by system $(\bar{F},\bar{G},\bar{H})$ , where the functions $\bar{F}_{i-1},\bar{G}_{i-1},\bar{H}_{i-1}$ only depend on $t_{x_0},\dots,t_{x_{i-1}}$ .For instance, $x_3=F_2(x_1,t_{x_1};x_2,t_{x_2})=$ $=F_2(F_0,t_{x_1};F_1(F_0,t_{x_1}),t_{x_2})=\bar{F}_2(t_{x_1},t_{x_2})$.

It is obvious that every $R$ given by $(F,G,H)$ can be represented by some sequence $L_1,\dots,L_N$ .However, if we require that only the condition (a) is satisfied, then there are rules $L_1,\dots,L_N$ which can not be represented by $(F,G,H)$ . In order to prove that, it is enough to consider an arbitrary sequence of non-constant functions $J_1,\dots,J_N$ ,defined on the set of binary sequences of lenght $N-1$ , $J_i\in\{0,1\}$ . Let $L_i(t_1,\dots,t_N)=J_i(t_1,\dots,t_{i-1},t_{i+1},\dots,t_N)$ .Then $L_i$ does not depend on $t_i$ .If a system $(F,G,H)$ exists, then $L_{F_0}=G_0=$const.

which contradicts the assumption that $L_{F_0}$ is a non-constant function.

<u>Definition</u> (Kolmogorov, [3]) A sequence $T$ is $(n,\varepsilon,p|R)$-random, $1\leq n\leq N$, $0<\varepsilon\leq 1$, $0\leq p\leq 1$, if $\nu<n$ or $\nu\geqslant n$, and $|f_1-p|<\varepsilon$. For systems $\mathcal{R}=\{R_1,R_2,\dots\}$, the sequence $T$ is $(n,\varepsilon,p|\mathcal{R})$-random if it is $(n,\varepsilon,p|R_i)$-random for each $i$.

For some $N,n,\varepsilon,p$ there exist sequences which are $(n,\varepsilon,p|R)$-random for each algorithm $R$.

Indeed, the sequence $T$ is $(n,\varepsilon,p|R)$-random for each $R$ if the following condition is satisfied: (*) for every $k$, $n\leq k\leq N$, and every subsequence $S$ of the lenght $k$, $|f_1-p|<\varepsilon$.

Then, $p-\min\{1,\frac{N-i}{n}\}\leq p-f_1\leq\min\{1,\frac{i}{n}\}-q$ holds, where $i$ is the number of zeros in $T$ and $q=1-p$. The condition (*) is satisfied if and only if $\min\{1,\frac{i}{n}\}-q<\varepsilon$ and $p-\min\{1,\frac{N-i}{n}\}>-\varepsilon$.

Let $p\geqslant\frac{1}{2}$. If one of the following conditions (i) $p<\varepsilon$ (ii) $p\geqslant\varepsilon$, $p>1-\varepsilon$ and $0\leq i<n(q+\varepsilon)$ (iii) $p\leq 1-\varepsilon$ and $N-n(p+\varepsilon)<i<n(q+\varepsilon)$ is satisfied, then the condition (*) is also satisfied. It is easy to see that there exists a sequence $T$ for which (iii) is satisfied if and only if $p\leq 1-\varepsilon$ and $n>\frac{N+\theta_n}{1+2\varepsilon}$, $\theta_n=n(q+\varepsilon)-[n(q+\varepsilon)]^+$, where $[x]^+$ is the largest integer being less then $x$.

Let us add this obvious statement: if a sequence $T$ has a subsequence of the lenght $\nu\geqslant n$ for which $|f_1-p|\geqslant\varepsilon$ then there exists an algorithm $R$ such that $T$ is $(n,\varepsilon,p|R)$-non random.

Let us show that for arbitrary set $\mathcal{T}$ of sequences $T$ which does not satisfy condition (*), there does not exist any algorithm $R$ such that $T$, $(T\in\mathcal{T})$ is $(n,\varepsilon,p|R)$-non-random.

Let $P$ be a probability measure on the set of all sequences of the length $N$. Let $P(n,\varepsilon,p|\mathcal{R})$ be $P$-measure of the set of sequences which are $(n,\varepsilon,p|R)$ -non-random. Let us put $P(n,\varepsilon,p) = \max_R P(n,\varepsilon,p|R)$. Let $P(n,\varepsilon,p) < 1$ and let $\mathcal{T}$ be such that $P(\mathcal{T}) > P(n,\varepsilon,p)$. This means that an $R$ does not exist for which each $T$, $(T \in \mathcal{T})$ is $(n,\varepsilon,p|R)$ -non-random. We have left to show that there exist $N,n,\varepsilon,p$ so that $P(n,\varepsilon,p)<1$ and that condition $(*)$ is not satisfied. In [3] for Bernoulli measure $P$ the estimation $P(n,\varepsilon,p) \leq 2e^{-2n\varepsilon^2(1-\varepsilon)}$ is given. It means that, if we choose $N,n,\varepsilon,p$ so that $\frac{\log 2}{2\varepsilon^2(1-\varepsilon)} < n \leq$ $\leq \frac{N+\theta_n}{1+2\varepsilon}$, $\frac{1}{2} \leq p \leq 1-\varepsilon$, then $P(n,\varepsilon,p) < 1$ and the condition $(*)$ is not satisfied. This shows that with this idea of randomness it is necessary to consider systems of algorithms contrary to the case of the complexity idea where a universal test for randomness exists. Actually, if $\mathcal{T}$ is the set of sequences with certain "bias" with respect to randomness, then, generaly, for their discovery one algorithm is not enough.

Let system $\mathcal{R}$ have $\rho$ algorithms. Let $\rho^*$ be such a number that each system with $\rho^*$ algorithms has at least one random sequence. Let us put $\sigma = \max \rho^*$. Some estimations of $\sigma$ are given in [3] and [1]. On the other side, let $\bar{\rho}$ be such a number that each system with $\bar{\rho}$ algorithms does not have any random sequence. Let us put $\tau = \min \bar{\rho}$. Since more than one different algorithm can reject the same sequences as non-random, then there will exist systems with large number of algorithms rejecting a small number of sequences. Hense, the number $\tau$ is not interesting in general case. For example, if $\varepsilon > \frac{1}{2N}$ then $\tau \geq N!$. Indeed, if $\varepsilon > \frac{1}{2N}$, then for each $p$ at least one sequence will exist such that $|f_1 - p| < \varepsilon$, where $f_1$ is relative frequency of the accurancy of the number one in the whole sequence. Let $R$ select the whole sequence. There are at least $N!$ of such algorithms, for it is enough that two algorithms differ for a function $F$. (In fact, there are exactly $N \cdot 2^{\frac{(N-2)(N+1)}{2}}$ different systems of functions $\bar{F}$.

Since there are more different algorithms then different sequences, instead of a arbitrary system, we consider minimal systems ( [1] ), i.e. the systems in which, for each algorithm, there exists at least one sequence rejected only by that algorithm. Let us put $\tau_0 = \min \bar{\rho}$ , where $\min$ is taken over all minimal systems. Note that some estimations for $\tau_0$ are given in [1] .

References:


[1] D.Banjević and Z.Ivković: On algorithmical testing tables of random numbers, Publication de l'Institut Mathematique, Belgrade, T.25, (1979), pp 11-15.

[2] A.Church: On the concept of a random sequence, Bull. Amer.Math. Soc. , 46, (194o), pp 13o-135.

[3] A.N.Kolmogorov: On tables of random numbers, Sankhya, ser. A 25, (1963), pp 369-376.

[4] А.К.Звонкин, Л.А.Левин: Сложность конечных обьектов и обоснование понятия информации с помощью теории алгоритмов, УМН, XXV, 6 (1970), 86-127.

[5] А.Н.Колмогоров: Три подхода к определению понятия "количество информации", Пробл. пер. информации, I, 1 (1965), 3-7.

[6] Л.А.Левин: О понятии случайной последовательности, Доклады АН СССР, т. 212, 3 (1973), 548-555.

# SOME REMARKS ON THE $BMO_\Phi$ SPACES.

N.L.BASSILY

University of Budapest

It is known that the $BMO_p$ spaces $(1<p<+\infty)$ are useful in the theory of analysis as well as in the theory of probability. Especially they play an important role in the theory of martingales.

Therefore, it is interesting to generalize the notion of these spaces and to obtain in such a way more general results than those obtained before. By using the Young functions we define the $BMO_\Phi$ space corresponding to the Young function $\Phi$.

These definitions and results will enable us to give more general forms with better constants of some interesting and useful inequalities. They also help us to show that for a great class of Young functions larger than the convex power ones, the $BMO_1$ - norm of a random variable is equivalent to its $BMO_\Phi$ norm.

To reach this generalization we have also introduced the notion of the conditional $L^\Phi$ -norm which can be considered as a norm, e.g, it satisfies a.e. the properties of an ordinary norm. It vainshes iff the random variable equals to $o$ a.e., positive homogeniety, and the triangle inequality are satisfied.

We have defined the conditional $L^{\Phi}$ -norm for a Young function $\Phi$ as follows: let $X$ be a random variable on the probability space $(\Omega, A, p)$, $F \subset A$, is a $\sigma$-field. The set of random variables $\gamma$ is defined by

$$F_X^{\Phi, F} = \{\gamma : \gamma > o \quad \text{a.e.}, \quad F\text{-measurable}, \quad E(\Phi(\frac{|X|}{\gamma}) \mid F) \leq 1 \text{ a.e.}\}.$$

We say that $X \in L_{\Phi}^{F}$ if $F_X^{\Phi, F}$ is not empty and in this case we define $\|X\|_{\Phi}^{F} = \text{ess. inf. } F_X^{\Phi, F}$

We define $\|X\|_{\Phi}^{F} = +\infty$, if $F_X^{\Phi, F}$ is empty.

The existence and uniqueness of $\|X\|_{\Phi}^{F}$ is out of question. To clarify this definition we have the following assertion.

Theorem (1):

If $X \in L^{\Phi}$, the Orlicz-space defined by the Young function $\Phi$, then $X \in L_{\Phi}^{F}$, where $F$ is an arbitrary $\sigma$-field such that $F \subset A$.

As a special case, if $\Phi(x) = x^{p}$, $1 < p < +\infty$, $X \in L^{p}$, then $F_X^{\Phi, F}$ is not empty and

$$\|X\|_{\Phi}^{F} = (E(|X|^{p} \mid F))^{\frac{1}{p}} .$$

The random variable $\|X\|_{\Phi}^{F}$ has the properties of a norm a.e.

Theorem (2):

Let $X \in L_{\Phi}^{F}$. Then,

a. For any real $c$, we have

$$cX\in L_{\Phi}^{F} \quad \text{and}, \quad \| cX\|_{\Phi}^{F} = |c| \ \| X \|_{\Phi}^{F} \quad \text{a.e.}$$

b. $\| X \|_{\Phi}^{F} = 0$ a.e. iff $X=0$ a.e.

c. If $Y\in L_{\Phi}^{F}$, then, $X+Y\in L_{\Phi}^{F}$, moreover,

$$\| X+Y \|_{\Phi}^{F} \leq \| X \|_{\Phi}^{F} + \| Y \|_{\Phi}^{F} \quad \text{a.e.}$$

The property a., is satisfied in a more general situation too. Namely, we have

a′. Let $Y\neq 0$ a.e. be $F$-measurable random variable, and suppose that $X\in L_{\Phi}^{F}$ as well as $XY\in L_{\Phi}^{F}$. Then,

$$\| YX \|_{\Phi}^{F} = |Y| \ \| X \|_{\Phi}^{F} \quad \text{a.e.}$$

Let $X\in L^{1}$ be a random variable and $(Fn)$, $n\geq 0$, an increasing sequence of $\sigma$-fields of events.
We suppose that $F_{\infty} = \sigma(\bigcup_{n=0}^{\infty} Fn) = A$.

Consider the martingale

$$X_{n} = E(X \mid Fn), \quad n\geq 0$$

We suppose that $X_{0}=0$ a.e.
Let $\Phi$ be a Young function. The $BMO_{\Phi}$ space is defined in the following way:

We say that $X\in BMO_{\Phi}$ if

$$\| \sup_{n\geq 1} \| X-X_{n-1} \|_{\Phi}^{Fn} \|_{\infty} < +\infty \ .$$

Especially if $\Phi(x)=x^p$ , $1<p<+\infty$, then,

$$\| X-X_{n-1} \|_{\Phi}^{Fn} = (E\,|X-X_{n-1}|^p \,|\, Fn))^{\frac{1}{p}}$$

So the above definition of $BMO_{\Phi}$ reduces to the well known $BMO_p$ space.

Concerning this definition we have the following results.

Theorem (3):

If $X \in BMO_{\Phi}$ , where $\Phi$ is a Young function, then

$$\| X \|_{BMO_{\Phi}} = \| \sup_{n \geq 1} \| X-X_{n-1} \|_{\Phi}^{Fn} \|_{\infty} \quad \text{is a norm.}$$

Theorem (4):

Suppose that $X \in BMO_{\Phi}$ , where $\Phi$ is a Young function. If $\Phi^*$ is another Young function such that

$$I = \int_0^{+\infty} \frac{\varphi^*(\lambda)}{\Phi(\lambda)} \, d\lambda < +\infty \,, \quad \text{then} \quad X \in BMO_{\Phi^*}$$

moreover, we have

$$\| X \|_{BMO_{\Phi^*}} \leq \max\,(1,I) \quad \| X \|_{BMO_{\Phi}} \quad .$$

Here $\varphi^*$ denotes the right-hand side derivative of $\Phi^*$ . Also, we have established the last assertion in another formulation to avoid the diffic lties concerning the integrability of $\frac{\varphi^*(\lambda)}{\Phi(\lambda)}$ in the neighbourhood of $\lambda=o$ as follows:

If for some $\lambda_o \geq o$ , the integral

$$I_o = \int_{\lambda_o}^{+\infty} \frac{\varphi^*(\lambda)}{\Phi(\lambda)} \, d\lambda \,, \quad \text{converges, then} \quad X \in BMO_{\Phi^*}$$

and $$\| X \|_{BMO_{\Phi^*}} \leq \max(1, I_o + \Phi^*(\lambda_o)) \; \| X \|_{BMO_\Phi} .$$

Supposing that the other conditions of theorem (4) are satisfied.

It is well known that when $X \in BMO_1$ then the random variables $\sup_{n \geq k} |X_n - X_{k-1}|$ are such that

$$E(e^{t \sup_{n \geq k} |X_n - X_{k-1}|} \mid F_k) \in L^\infty \quad \text{with a}$$

suitable $t > o$ for all $k \geq 1$, (Garsia [1], Theorem III.2.2.)

and consequently, $$E(e^{t |X - X_{k-1}|} \mid F_k) \in L^\infty$$

This fact is very naturally embedded in our notion of $BMO_\Phi$ spaces with $\Phi(x) = e^x - x - 1$ in the following theorem.

Theorem (5):

If $X \in BMO_1$, then $X \in BMO_\Phi$, where $\Phi$ is the Young function $\Phi(x) = e^x - x - 1$, moreover,

$$\| X \|_{BMO_\Phi} \leq 8 \; \| X \|_{BMO_1} .$$

We know that when $X \in BMO_1$, then $X \in BMO_p$, $1 < p < +\infty$. Theorem (6) below will extend the validity of this fact to an essentially larger class of Young functions more than the convex power ones.

Theorem (6):

Let $X \in BMO_1$, then $X \in BMO_\Phi$, where $\Phi$ is any Young function for which

$$I = \int_{o}^{+\infty} \frac{\varphi(\lambda)}{e^{\lambda} - \lambda - 1} d\lambda \qquad \text{converges.}$$

In this case we also have

$$\| X \|_{BMO_{\Phi}} \leq 8 \max(1, I) \; \| X \|_{BMO_1}$$

If for some $\lambda_o \geq o$ the integral

$$I_o = \int_{\lambda_o}^{+\infty} \frac{\varphi(\lambda)}{e^{\lambda} - \lambda - 1} d\lambda$$

converges, then $X \in BMO_1$ implies that $X \in BMO_{\Phi}$, and

$$\| X \|_{BMO_{\Phi}} \leq 8 \max(1, I_o + \Phi(\lambda_o)) \; \| X \|_{BMO_1} \, .$$

Especially, for any Young-function $\Phi$ for which the corresponding power $p = \sup\limits_{x>o} \frac{x\varphi(x)}{\Phi(x)}$ is finite, there is $\lambda_o$ such that the corresponding integral $I_o$ converges.

Also, we have the following corollary,

Corollary:

If $X \in BMO_{\Phi}$, where $\Phi$ has a finite power $p$, then we have:

$$c_{\Phi} \| X \|_{BMO_1} \leq \| X \|_{BMO_{\Phi}} \leq C_{\Phi} \; \| X \|_{BMO_1} \, .$$

where $c_{\Phi}$ and $C_{\Phi}$ are two constants depending only on $\Phi$.

## REFERENCES

[1] GARSIA, A.M.: Martingale inequalities. Benjamin readings, Massachusetts, 1973.

[2] NEVEU, J.: Discrete parameter martingales. North-Holland, Amsterdam 1975.

# ON UNBIASED ESTIMATION OF A COMMON MEAN OF TWO NORMAL DISTRIBUTIONS

B. Bednarek-Kozek

University of Wrocław

Let $X_1, X_2, \ldots, X_n, Y_1, Y_2, \ldots, Y_m$ be independent normal random variables with $EX_i = EY_j = \theta$, $\theta \in R$, $\mathrm{Var}\, X_i = \sigma_1^2$, $\mathrm{Var}\, Y_j = \sigma_2^2$, $\sigma_1 > 0$, $\sigma_2 > 0$, $i=1,\ldots,n$, $j=1,\ldots,m$, $n > 1$, $m > 1$.

It is well known that there exists no unbiased uniformly minimum variance estimator for $\theta$ (A.M.Kagan [2]). Recent investigations of a structure of the class of estimators uniformly best unbiased for convex loss functions (B.Bednarek-Kozek, A.Kozek [1], A.Kozek [3]) do not exclude, however, the existence of a natural strictly convex loss function $L(d-\theta)$ which admits the existence of uniformly best unbiased estimator for $\theta$. In this paper we prove that there do not exist a smooth strictly convex loss function $L(d-\theta)$ and a smooth estimator X such that X could be uniformly best (with respect to $L(d-\theta)$) in the class of unbiased estimators for $\theta$. In the proof of this result we use the Wijsman's D-method (R.A.Wijsman [4]) originally used for testing problems.

In the considered problem the statistics $T_1 = \sum_{i=1}^{n} X_i^2$, $T_2 = \sum_{i=1}^{n} X_i$, $T_3 = \sum_{j=1}^{m} Y_j^2$, $T_4 = \sum_{j=1}^{m} Y_j$ are sufficient and generate the minimal sufficient $\sigma$-algebra S. The sufficient statistic $T=(T_1,T_2,T_3,T_4)$ has an exponential distribution with the density with respect to the Lebesgue measure $\mu$ in $R^4$ given by

$$c(\theta, \sigma_1, \sigma_2)h(t)\exp\left(\sum_{i=1}^{4} s_i t_i\right) ,$$

where $s_1(\sigma_1) = \frac{-1}{2\sigma_1^2}$, $s_2(\theta,\sigma_1) = \frac{\theta}{\sigma_1^2}$, $s_3(\sigma_2) = \frac{-1}{2\sigma_2^2}$, $s_4(\theta,\sigma_2) = \frac{\theta}{\sigma_2^2}$

and

$$h(t) = \begin{cases} (nt_1-t_2^2)^{(n-3)/2}(mt_3-t_4^2)^{(m-3)/2} & \text{if } nt_1 \geq t_2^2,\ mt_3 \geq t_4^2 \\ 0 & \text{otherwise} \end{cases}$$

Assume that the considered set $\mathcal{E}$ of estimators is a vector space containing all bounded and Lebesgue measurable estimators. Denote by $\mathcal{E}_0$ the vector space of all unbiased estimators of zero in $\mathcal{E}$, i.e., $X \in \mathcal{E}_0$ if and only if $X \in \mathcal{E}$ and $E_{\theta,\sigma_1,\sigma_2}X = 0$ for each $\theta \in R$, $\sigma_1 > 0$, $\sigma_2 > 0$. The set of parameters

$$\Omega = \{(s_1(\sigma_1), s_2(\theta,\sigma_1), s_3(\sigma_2), s_4(\theta,\sigma_2));\ \theta \in R,\ \sigma_1 > 0,\ \sigma_2 > 0\}$$

is the subset of points $s = (s_1, s_2, s_3, s_4) \in R^4$ satisfying the relations:

(1) $$s_1 s_4 - s_2 s_3 = 0,\quad s_1 < 0,\quad s_3 < 0 .$$

Now we formulate and prove a theorem somewhat more general, then the result announced at the begining of the paper.

Theorem. Let $a = a(s_1, s_2)$ ($\neq$ const) be a parametric function and let $L(d-a)$ be a loss function, where $L$ is convex, non-negative and has the third derivative continuous. If $X \in \mathcal{E}$ has continuous second partial derivatives and there exists $\varepsilon > 0$ such that $E_s L(\lambda X(T) - a) < \infty$ for each $s \in \Omega$ and $\lambda \in (1-\varepsilon, 1+\varepsilon)$, then $X$ is not uniformly best unbiased for $a$.

Corollary. If $a = \theta = -\frac{s_2}{2s_1}$ and if the assumptions of the Theorem on $L$ and $X$ are fulfiled, then $X$ is not uniformly best unbiased for $\theta$.

Proof of the Theorem. The proof bases on the Lehmann-Scheffe-Rao Lemma. So, we shall need a class of unbiased estimators of zero.

First note that if $g$ is a function with a continuous second partial derivatives and with a compact support contained in the interior of the support of $h$, then the integration by parts yields

$$E_s \left\{ \left( \frac{\partial^2 g(t)}{\partial t_1 \partial t_4} - \frac{\partial^2 g(t)}{\partial t_2 \partial t_3} \right) \frac{1}{h(t)} \right\} = (s_1 s_4 - s_2 s_3)\, E_s\left( \frac{g(t)}{h(t)} \right) = 0 .$$

Hence if $z(t)$ is given by

$$z(t) = \begin{cases} \left( \frac{\partial^2 g(t)}{\partial t_1 \partial t_4} - \frac{\partial^2 g(t)}{\partial t_2 \partial t_3} \right) \frac{1}{h(t)} & \text{if } h(t) > 0 \\ 0 & \text{if } h(t) = 0 \end{cases}$$

then $z(T)$ is an unbiased estimator of zero. Let

$$\mathcal{C} = \left\{ z(T) \; : \; z(t) = \left( \frac{\partial^2 g(t)}{\partial t_1 \partial t_4} - \frac{\partial^2 g(t)}{\partial t_2 \partial t_3} \right) \frac{1}{h(t)} \quad , \; g(\cdot) \in G \right\}$$

where G is the class of all functions with continuous second partial derivatives and with compact supports contained in the interior of the support of h . Clearly $\mathcal{C} \subset \mathcal{E}_0$ .

Suppose that $x(T)$ is a uniformly best unbiased estimator of $a = a(s_1, s_2)$ and $E_s L(\lambda x(T) - a)$ is finite for $\lambda \in (1-\varepsilon, 1+\varepsilon)$ and $s \in \Omega$ . Then $x(T)$ is uniformly best unbiased for a in the class $\{x(T) + z(T) \; ; \; z(T) \in \mathcal{C}\}$ and it is easily seen that

$$E_s L(x(T) + \gamma z(T) - a) < \infty \qquad \text{for } |\gamma| < \frac{\varepsilon}{1+\varepsilon} \quad .$$

Thus by the Lehmann-Scheffe-Rao Lemma ( [3] Appendix A )

$$E_s L'(x(T) - a) z(T) = 0$$

holds for each $z(T) \in \mathcal{C}$ .

Hence we get

$$\int_{R^4} \exp\left( \sum_{i=1}^{4} s_i t_i \right) L'(x(t) - a) \left( \frac{\partial^2 g(t)}{\partial t_1 \partial t_4} - \frac{\partial^2 g(t)}{\partial t_2 \partial t_3} \right) dt = 0 \tag{2}$$

for every $s \in \Omega$ and $g \in G$ . Integration by parts of the integral on the left-hand side of (2) yields

$$\int_{R^4} \left( \frac{\partial^2}{\partial t_1 \partial t_4} - \frac{\partial^2}{\partial t_2 \partial t_3} \right) \left\{ \exp\left( \sum_{i=1}^{4} s_i t_i \right) L'(x(t) - a) \right\} g(t) dt = 0$$

for every $s \in \Omega$ and $g \in G$ . Hence

$$\left( \frac{\partial^2}{\partial t_1 \partial t_4} - \frac{\partial^2}{\partial t_2 \partial t_3} \right) \left\{ \exp\left( \sum_{i=1}^{4} s_i t_i \right) L'(x(t) - a) \right\} = 0 \tag{3}$$

holds for every $s \in \Omega$ and $t \in \text{int supp } h$ . It is easy to see that (3) is equivalent to the equality

(4)
$$\left(s_1\frac{\partial x}{\partial t_4} + s_4\frac{\partial x}{\partial t_1} - s_2\frac{\partial x}{\partial t_3} - s_3\frac{\partial x}{\partial t_2}\right) L''(x(t)-a) +$$

$$\left(\frac{\partial x}{\partial t_1}\frac{\partial x}{\partial t_4} - \frac{\partial x}{\partial t_3}\frac{\partial x}{\partial t_2}\right) L'''(x(t)-a) + \left(\frac{\partial^2 x}{\partial t_1 \partial t_4} - \frac{\partial^2 x}{\partial t_2 \partial t_3}\right) L''(x(t)-a) = 0$$

for $s \in \Omega$ and $t \in \text{int supp } h$ .

Since , in view of (1) we have $s_4 = \frac{s_2 s_3}{s_1}$ for $s \in \Omega$, equality (4) can be rewritten in the following form

(5)
$$s_3\left(\frac{s_2}{s_1}\frac{\partial x}{\partial t_1} - \frac{\partial x}{\partial t_2}\right)L''(x(t)-a) + \left(s_1\frac{\partial x}{\partial t_4} - s_2\frac{\partial x}{\partial t_3}\right)L''(x(t)-a) +$$

$$\left(\frac{\partial x}{\partial t_1}\frac{\partial x}{\partial t_4} - \frac{\partial x}{\partial t_2}\frac{\partial x}{\partial t_3}\right)L'''(x(t)-a) + \left(\frac{\partial^2 x}{\partial t_1 \partial t_4} - \frac{\partial^2 x}{\partial t_2 \partial t_3}\right)L''(x(t)-a) = 0,$$

$(s_1, s_2, s_3) \in R^- \times R \times R^-$ , $t \in \text{int supp } h$ . The second , third and fourth components of the left-hand side of (5) do not depend on $s_3$ .Thus, the strict convexity of L implies

(6)
$$\frac{s_2}{s_1}\frac{\partial x}{\partial t_1} - \frac{\partial x}{\partial t_2} = 0$$

for every $s_1 \in R^-$ , $s_2 \in R$ , $t \in \text{int supp } h$ . Hence $\frac{\partial x}{\partial t_1} = \frac{\partial x}{\partial t_2} = 0$ holds and (5) implies that $\frac{\partial x}{\partial t_3} = \frac{\partial x}{\partial t_4} = 0$ for $t \in \text{int supp } h$ . Thus $x(t)=\text{const}$ with probability 1. This contradicts our assumption that $E_s x(T) = a(s_1, s_2) \neq \text{const}$ . The proof of the Theorem is completed.

REFERENCES

[1] B.Bednarek-Kozek, A.Kozek, Two examples of strictly convex non-universal loss functions. Institute of Mathematics , Polish Academy of Sciences, (1978) Preprint No 133.

[2] A.M.Kagan. Estimation theory for families with location and scale parameters, and for exponential families. Proc. Steklov.Inst.Math.104 (1968) 21 – 103, Trudy Mat.Inst. Steklov 104 (1968).

[3] A.Kozek. On two necessary $\sigma$-fields and on universal loss functions. Institute of Mathematics, Polish Academy of Sciences, (1979) Preprint No 174 (to appear in Probability and Mathematical Statistics vol.1 ).

[4] R.A.Wijsman. Incomplete sufficient statistics and similar tests. Ann.Math.Statist.29 (1958) 1028 – 1045.

# THE INVARIANCE PRINCIPLE FOR VECTOR VALUED RANDOM VARIABLES WITH APPLICATIONS TO FUNCTIONAL RANDOM LIMIT THEOREMS

T. BYCZKOWSKI and T. INGLOT

<u>1. Introduction.</u> Let $\xi_1, \xi_2, \ldots$ be a sequence of independent real valued random variables with the same distribution with mean 0 and variance 1. Define the stochastic process $\zeta_n$ by

$$\zeta_n(t) = S_{[nt]}/n^{1/2} = (\xi_1 + \ldots + \xi_{nt})/n^{1/2},$$

where $0 \leq t \leq 1$ and $n = 1, 2, \ldots$ The classical Invariance Principle states that the sequence of $D[0,1]$ valued random variables induced by the processes $\zeta_n(t)$ converges weakly to the Wiener process.

The first generalization of the result of this type to (separable) Banach space valued random variables was given by Kuelbs [12]. The case of random variables taking values in complete separable metric linear spaces was investigated in [2] and [11]. However, results obtained there were not completely satisfactory. First of all, these extensions are valid only for random variables taking values in a relatively narrow class of vector spaces (such as locally pseudoconvex spaces or spaces of measurable functions).

On the other hand, the investigation of stochastic processes without discontinuities of second kind yields several interesting problems concerning random variables with values in the space $D[0,1]$, which is no longer a topological vector space (addition is not continuous). Therefore, it seems to be desirable to extend the Invariance Principle to random variables with values in more general vector spaces (containing both the complete separable metric vector spaces and the space $D[0,1]$).

Such an extension is given in Section 3. Our approach is based on the paper [2]. In Section 4 we discuss the generalization of the classical method of random change of time (see § 17 in [1]). Arguments used there are slight modifications of classical ones, so we restrict ourselves to point out only a few differences between the classical and the general cases. In Section 5 we apply these results to obtain several functional limit theorems. Results obtained there are, in most cases, known. Our approach, however, allows us to derive all these results as simple consequences of those in Section 3. The authors hope that this method will provide some unification.

2. Preliminaries. Let $(E,\varrho)$ be a complete separable metric space endowed with a fixed complete metric $\varrho$. By $\mathcal{E}$ we denote the $\sigma$-algebra of Borel subsets of E. Let $D_E = D_E[0,1]$ be the space of all functions defined on the unit interval into E that are right continuous and have left-hand limits. We quote here some statements concerning the metric and measure-theoretic properties of the space $D_E$. All these statements are well known when E is the real line. The proofs, however, are also valid in our more general situation; the interested reader is referred to [1].

Let $\Lambda$ denote the class of strictly increasing, continuous mappings $\lambda$ of $[0,1]$ onto itself taking 0 to 0 such that $\|\lambda\| < \infty$, where

$$|\lambda| = \sup_{s \neq t} \left| \ln \frac{\lambda s - \lambda t}{s-t} \right| .$$

For f and g in $D_E$ define $d_o(f,g)$ to be the infimum of those positive $\varepsilon$ for which

$$\sup_t \rho(f(t), g(\lambda t)) < \varepsilon$$

with some $\lambda \in \Lambda$ satisfying $|\lambda| < \varepsilon$. $d_o$ is a metric and $D_E$ endowed with $d_o$ is a complete separable metric space.

For $t_1, \ldots, t_k$, $0 \leq t_i \leq 1$, $i=1,\ldots,k$, define the natural projection $\pi_{t_1 \ldots t_k}$ from $D_E$ to $E^k$:

$$\pi_{t_1 \ldots t_k}(f) = (f(t_1), \ldots, f(t_k)) .$$

These projections are Borel measurable and generate the $\sigma$-algebra of Borel subsets of $D_E$.

A mapping X from a probability space $(\Omega, \sigma, P)$ into $D_E$ (E resp.) will be called a $D_E$ (E resp.) valued random variable if it is measurable with respect to the $\sigma$-algebra of Borel subsets of $D_E$ (E resp.). X is a $D_E$ valued random variable if and only if for every $t \in [0,1]$ $X(t) = \pi_t(X)$ is an E valued random variable and for almost all $\omega \in \Omega$ $t \to X(t,\omega)$ belongs to $D_E$.

Now, we state a sufficient condition of convergence in distribution of $D_E$ valued random variables which will be used in the sequel (see § 15 in [1]).

Proposition 1. Let $(X_n)_{n=1}^{\infty}$ be a sequence of $D_E$ valued random variables satisfying the following conditions:

(i) for every $k \geq 1$ and arbitrary $t_1, \ldots, t_k$, $0 \leq t_i \leq 1$, the sequence $(X_n(t_1), \ldots, X_n(t_k))$ converges in distribution on $E^k$,

(ii) for each $\varepsilon, \eta > 0$ there exists $\delta > 0$ such that for every $n \geq 1$

$$P\{\sup_{|t-s|<\delta} \varrho(X_n(t), X_n(s)) \geq \varepsilon\} \leq \eta \quad .$$

Then the sequence $X_n$ converges in distribution to a $D_E$ valued random variable X such that $P\{X \in C_E\} = 1$, where $C_E$ denotes the space of all E valued continuous functions on $[0,1]$.

3. The Invariance Principle. Let $(E,\varrho)$ be a complete separable metric space with a fixed complete metric $\varrho$. Throughout the remaining part of this paper we will assume that E is a real vector space such that the addition is $\mathcal{E}\times\mathcal{E}$ measurable and the multiplication by scalars is jointly continuous. We will also assume that the metric $\varrho$ has the following quasi-subinvariance property:

$$\varrho(x+y,y) \leq \varrho(x,0)$$

for every $x,y \in E$. Let us denote

$$\|x\| = \varrho(x,0) .$$

It is evident that $\|x+y\| \leq \|x\|+\|y\|$.

Observe that separable Frechet spaces with invariant (and hence complete) metric as well as the space $D_R = D[0,1]$ with the metric $d_o$ satisfy all the assumptions imposed on $(E,\varrho)$ .

An E valued random variable $\xi$ will be called symmetric Gaussian (in the sense of Fernique [8]) if for every pair $(\xi_1, \xi_2)$ of independent copies of $\xi$ and for every pair of real numbers $(s,t)$ such that $s^2+t^2 = 1$ the random variables

$$s\xi_1 + t\xi_2 \quad \text{and} \quad t\xi_1 - s\xi_2$$

are independent and have the distribution of $\xi$ .

Given an E valued symmetric Gaussian random variable $\xi$ define $\mu_t$ to be the distribution of $t^{1/2}\xi$ , for every $0 \leq t \leq 1$. Then $(\mu_t)_{t\geq 0}$ is a convolution semigroup (that is, $\mu_t * \mu_s = \mu_{t+s}$)

of symmetric Gaussian probability measures. Moreover, this semigroup is continuous: $\mu_t$ converges weakly to $\delta_0$ when t tends to 0. Observe that by the Kolmogorov's Extension Theorem there exists an E valued homogeneous stochastic process $W_\xi(t)$ with independent increments such that $W_\xi(t)$ has the distribution $\mu_t$. If $W_\xi$ has continuous sample paths with probability 1 then it will be called the E valued Wiener process generated by $\xi$ .

Let $(\xi_n)_{n=1}^{\infty}$ be a sequence of independent, identically distributed E valued random variables defined on a common probability space. Define $S_0 = 0$, $S_n = \xi_1+\ldots+\xi_n$ and $X_n(t) = S_{[nt]}/ n^{1/2}$. Obviously, $X_n$ is a sequence of $D_E$ valued random variables. Our aim is to find conditions under which $X_n$ converges in distribution to an E valued Wiener process.

<u>Theorem 1.</u> Let $(\xi_n)_{n=1}^{\infty}$ be a sequence of independent identically distributed E valued random variables. Assume that $S_n/n^{1/2} = (\xi_1+\ldots+\xi_n)/n^{1/2}$ converges weakly to a symmetric E valued Gaussian random variable $\xi$ . Then the sequence $X_n$ of $D_E$ valued random variables, defined by $X_n(t) = S_{[nt]}/n^{1/2}$, converges in distribution to the Wiener process $W_\xi$ .

<u>Proof.</u> The idea of the proof is based on [2].

Using Proposition 1 and the same arguments as in the case E = R (see § 16 in [1]) we see that it suffices to verify the following condition:

for every $\varepsilon, \eta > 0$ there exist $\delta > 0$ and $n_0 \geq 1$ such that

$$\sup_{n\geq n_0} \sup_{k\leq n} \frac{1}{\delta} P\{ \sup_{i\leq n\delta} \varrho(S_{k-i}/n^{1/2}, S_k/n^{1/2}) \geq \varepsilon \} \leq \eta .$$

Using the quasi-subinvariance of $\varrho$ we see that this condition is satisfied whenever for every $\varepsilon, \eta > 0$ there exist $\delta > 0$ and $n_0 \geq 1$ such that

$$\sup_{n \geq n_o} \frac{1}{\delta} P\{ \sup_{i \leq [n\delta]} \| S_i/n^{1/2} \| \geq \varepsilon \} \leq \eta .$$

Now, observe that the above condition is satisfied whenever for every $\varepsilon, \eta > 0$ there exist $\delta > 0$ and $n_o \geq 1$ such that the following hold:

(A) $$\sup_{n \geq n_o} \frac{1}{\delta} P\{ \| S_{[n\delta]} /n^{1/2} \| \geq \varepsilon/2 \} \leq \eta/2$$

(B) $$\sup_{n \geq n_o} \sup_{i \leq n\delta} P\{ \| S_i/n^{1/2} \| \geq \varepsilon/2 \} \leq 1/2.$$

Indeed, if (A) and (B) hold then our condition is an immediate consequence of Ottaviani's Inequality (see, e.g., Lemma 4.1 in [13]):

$$\min_{1 \leq k \leq n} P\{ \| S_n - S_k \| \leq \varepsilon \} \; P\{ \max_{1 \leq k \leq n} \| S_k \| > \varepsilon \} \leq P\{ \| S_n \| > \varepsilon/2 \} .$$

That (B) is satisfied follows from the tightness of $S_n/n^{1/2}$ and the following consequence of the continuity of multiplication by scalars:

For every compact subset $K \subset E$ and $\varepsilon > 0$ there exists $h > 0$ such that for every $t \leq h^{1/2}$ $tK \subset \{x: \|x\| < \varepsilon\}$ .

To end the proof note that (A) follows from the convergence in distribution of $S_{[n\delta]}/n^{1/2}$ to $\delta^{1/2}\xi$ and from the following property of E valued symmetric Gaussian random variables (see Theorem 3.1 in [5]):

For every $\varepsilon > 0$ $\quad n P\{ \| \xi/n^{1/2} \| \geq \varepsilon \} \longrightarrow 0$ as $n \longrightarrow \infty$ .

Now, let $L_\Phi = L_\Phi(T, \mathcal{F}, m)$ be a separable Orlicz space, where $(T, \mathcal{F}, m)$ is a $\sigma$-finite measure space and $\Phi$ is a nondecreasing continuous nonnegative function defined on the positive half-line satisfying $(\Delta_2)$ condition and such that $\Phi(t) = 0$ if and only if $t = 0$.

In [3] it was shown that $L_\Phi$ valued random variables can be identified with measurable stochastic processes with sample paths in $L_\Phi$ . Thus, our theorem implies the following:

Corollary 1. Let $(\xi_n)_{n=1}^\infty$ be a sequence of independent identically distributed measurable stochastic processes with sample paths in $L_\Phi$ . Assume that $S_n/n^{1/2}$ converges in distribution on $L_\Phi$ to a Gaussian process $\xi$ with mean zero. Then the sequence of $D_{L_\Phi}$ valued random variables $X_n(t) = S_{[nt]}/n^{1/2}$ converges in distribution to the Wiener process $W_\xi$ .

Let us also mention the following consequence of our theorem and of Theorem 2 in [10] for stochastic processes with sample paths in D:

Corollary 2. Let $\xi$ be a stochastic process with sample paths in D satisfying $E\,\xi(t) = 0$, $E\,\xi^2(t) < \infty$, for every $t \in [0,1]$. Assume that there exist nondecreasing continuous functions G and F on $[0,1]$ and numbers $\alpha > 1/2$, $\beta > 1$ such that for all $0 \leq s \leq t \leq u \leq 1$ the following two conditions hold:

(i) $E\,(\xi(u) - \xi(t))^2 \leq (G(u) - G(t))^\alpha$ ,

(ii) $E\,(\xi(u) - \xi(t))^2\,(\xi(t) - \xi(s))^2 \leq (F(u) - F(s))^\beta$ .

If $(\xi_n)_{n=1}^\infty$ is a sequence of independent stochastic processes with sample paths in D, with the same distribution as $\xi$ , then $X_n(t)$ converges in distribution to a D valued Wiener process.

4. Random Change of Time. Let $(\xi_n)_{n=1}^\infty$ be a sequence of E valued random variables defined on a common probability space $(\Omega, \mathfrak{S}, P)$ and let $S_n$ and $X_n$ be defined as in Section 3. By $\nu_n$ we denote a sequence of integer-valued random variables defined on $(\Omega, \mathfrak{S}, P)$ and by $a_n$ - a sequence of positive reals tending to infinity. Define

$$Y_n = \frac{1}{\sqrt{\nu_n}} S_{[\nu_n t]} , \qquad Z_n = \frac{1}{\sqrt{a_n}} S_{[\nu_n t]} .$$

The following theorems are the vector valued versions of Theorems 17.1 and 17.2 in [1] .

Theorem 2. Assume that $\nu_n/a_n$ converges in probability to a positive constant $\theta$ and that $X_n$ converges in distribution to an E valued Wiener process W. Then

$$Y_n \xrightarrow{D} W \quad \text{and} \quad Z_n \xrightarrow{D} \theta^{1/2} W.$$

Proof. Our theorem can be proved essentially in the same way as Theorem 17.1 in [1]. Observe that exactly as in the proof of that theorem we can obtain that $Z_n$ converges in distribution to $W\circ\varphi$ , where $(W\circ\varphi)(t) = W(\varphi(t))$ and $\varphi(t) = \theta t$. Since W(t) has the same distribution as $t^{1/2} W(1)$ , $W\circ\varphi$ has the distribution of $\theta^{1/2}$ W. Since $Y_n = \sqrt{a_n/\nu_n}\ Z_n$, we obtain the desired conclusion.

Theorem 3. Assume that $\xi_1,\ldots, \xi_n,\ldots$ are independent and identically distributed and that $S_n/n^{1/2}$ converges in distribution to an E valued symmetric Gaussian random variable $\xi$ . If $\nu_n/a_n$ converges in probability to a positive random variable $\theta$ then

$$Y_n \xrightarrow{D} W_\xi \quad \text{and} \quad Z_n \xrightarrow{D} \theta_0^{1/2} W_\xi ,$$

where $\theta_0$ is independent of $W_\xi$ and has the same distribution as $\theta$ .

Proof. As in the proof of Theorem 17.2 in [1] we define a $D_E$ valued random variable $X_n'(t)$ by

$$X_n'(t) = \begin{cases} 1/n^{1/2} \sum\limits_{p_n\leq i\leq nt} \xi_i & \text{if } p_n \leq [nt] \\ 0 & \text{otherwise,} \end{cases}$$

where $p_n$ is a sequence of positive integers tending to infinity in such a way that $p_n/n \to 0$. Observe that

$$\delta_n = \sup_t \rho\,(X_n(t), X_n'(t)) \leq \sup_t \| (1/n)^{1/2} \sum_{i=1}^{\min(p_n-1,[nt])} \xi_i \| \leq$$

$$\leq \max_{1\leq k\leq p_n} \| (1/n)^{1/2} \sum_{i=1}^{k} \xi_i \| .$$

From the first part of the proof of Theorem 1 we infer that for every $\varepsilon > 0$ there exist $\delta > 0$ and $n_o$ such that for $n \geqslant n_o$ $\max_{k \leqslant n\delta} P\{\|(1/n)^{1/2} \sum_{i=1}^{k} \xi_i\| > \varepsilon/2\} < 1/2$. Next, by Ottaviani's Inequality we obtain

$$P\{\max_{1 \leqslant k \leqslant p_n} \|(1/n)^{1/2} \sum_{i=1}^{k} \xi_i\| > \varepsilon\} \leqslant 2\, P\{\|(1/n)^{1/2} \sum_{i=1}^{p_n} \xi_i\| > \varepsilon/2\} =$$

$$= 2\, P\{\|(p_n/n)^{1/2} \frac{1}{\sqrt{p_n}} \sum_{i=1}^{p_n} \xi_i\| > \varepsilon/2\}$$

for $n \geqslant n_o$ and such that $p_n \leqslant n\delta$. By the joint continuity of multiplication by scalars along with the fact that $p_n/n \to 0$ and $(1/p_n)^{1/2} \sum_{i=1}^{p_n} \xi_i \xrightarrow{D} \xi$, the last term tends to 0. Hence $\delta_n \to 0$ in probability. The remaining part of the proof can be derived in the same way as in Theorem 17.2 in [1].

## 5. Applications.

<u>(I) Random Central Limit Theorem.</u> Let $(\xi_n)_{n=1}^{\infty}$ be a sequence of E valued random variables satisfying the assumptions of Theorem 2 or 3. If $\nu_n$, $a_n$, $\theta_o$, $\xi$ are as in Section 4 then we have

$$\frac{1}{\sqrt{\nu_n}} S_{\nu_n} \xrightarrow{D} \xi, \quad \frac{1}{\sqrt{a_n}} S_{\nu_n} \xrightarrow{D} \theta_o^{1/2} \xi .$$

This theorem is a particular case of theorems obtained by several authors, among them by Prakasa Rao [14] and Fernandez [7], who investigated random variables with values in Hilbert spaces or normed spaces, respectively.

For E = R this theorem was obtained (in a more general setting) by Gnedenko - Fahim [9].

<u>(II) Random Mixture.</u> Let $\xi_1, \ldots, \xi_m$ be independent E valued random variables. Assume that for each i, $1 \leqslant i \leqslant m$, $\xi_i$ satisfies the CLT, that is, $(\xi_{i1} + \ldots + \xi_{in})/n^{1/2}$ converges in distribution to an E valued symmetric Gaussian random variable, where $(\xi_{ik})_{k=1}^{\infty}$ is a sequence of independent copies of $\xi_i$.

Let $(\delta_1, \ldots, \delta_m)$ be a $R^m$ valued random variable such that every $\delta_i$ takes values 0 or 1 only, and $\delta_1+\dots+\delta_m=1$. Define a random "mixture" of $\xi_i$ by

$$\xi = \sum_{i=1}^{m} \delta_i \xi_i .$$

In the investigation of limit properties of renewal or breakdown processes it is useful to know whether $\xi$ satisfies the CLT. We outline here the connection of this problem with the previous one, for details the reader is referred to [16].

Let $(\xi_{ik})_{k=1}^{\infty}$, $(\delta_{ik})_{k=1}^{\infty}$, $i = 1,\dots,m$, be sequences of independent copies of $\xi_i$ and $\delta_i$, respectively. Then

$$\xi^{(k)} = \sum_{i=1}^{m} \delta_{ik} \xi_{ik}$$

is a sequence of independent copies of $\xi$. Define

$$S_n = \sum_{k=1}^{n} \xi^{(k)} = \sum_{k=1}^{n} \sum_{i=1}^{m} \delta_{ik} \xi_{ik}.$$

It can be easily checked that $S_n$ has the same distribution as $S_n'$:

$$S_n' = \sum_{i=1}^{m} \sum_{k=1}^{\nu_{in}} \xi_{ik},$$

where $(\nu_{1n},\dots,\nu_{mn})$ is a $R^m$ valued random variable having the multinomial distribution. Assume additionally that either E is a topological vector space or E = D and all limiting Gaussian distributions are concentrated on C. The application of Theorem 1 and 2 shows that $S_n/n^{1/2}$ converges in distribution to an E valued symmetric Gaussian random variable.

III Random Sequences of Empirical Processes. Let $E = D[0,1] = D$. Let $\tau_1, \tau_2,\dots$ be a sequence of independent identically distributed real random variables such that $0 \leq \tau_n \leq 1$, $n = 1,2,\dots$ We write F for the distribution function of $\tau_i$. Define the sequence of D valued random variables $\xi_n$ by

$$\xi_n(t) = 1_{\{\tau_n \le t\}} - F(t).$$

$\xi_n$ is a sequence of independent, identically distributed D valued random variables and

$$\frac{1}{n} \sum_{i=1}^{n} \xi_i(t) = F_n(t) - F(t),$$

where $F_n(t)$ is the empirical distribution function of $(\tau_n)_{n=1}^{\infty}$. Theorem 16.4 in [1] states that

$$n^{1/2}(F_n - F) = 1/n^{1/2} \sum_{i=1}^{n} \xi_i \xrightarrow{D} W_o \circ F,$$

where $W_o$ is the Brownian Bridge and $W_o \circ F(t) = W_o(F(t))$. Theorem 3 yields now the following result:

If $\nu_n$ are as in Theorem 3 then

$$\frac{1}{\sqrt{\nu_n}} \sum_{i=1}^{\nu_n} \xi_i = \nu_n^{1/2}(F_{\nu_n} - F) \xrightarrow{D} W_o \circ F.$$

A particular case of this result $(\Theta = 1)$ was obtained by Pyke [15]. The general case was proved by Csörgő [6] (by a method completely different from ours). This result provides random limit theorems for Kolmogorov, Smirnov, Cramer - von Mises and Renyi statistics (see [6] for other applications).

Remark. We can also define an E valued Gaussian random variable $\xi$ as stable of index 2 and such that for every pair $(\xi_1, \xi_2)$ of independent copies of $\xi$ and every reals (s, t) such that $s^2+t^2=1$, the random variables $s\xi_1 + t\xi_2$ and $t\xi_1 - s\xi_2$ are independent. However, it follows from Corollary 3.2 in [4] that every E valued random variable in this sense is a translation of a symmetric one. Hence, all results stated in Theorems 1 - 3 and in (I) and (II) remain valid, under suitable modification, when the limiting Gaussian distribution is not necessarily symmetric.

## References.

[1] P. Billingsley, Convergence of probability measures, Wiley, 1967.

[2] T. Byczkowski, The invariance principle for group valued random variables, Studia Math. 56 (1976), 187-198.

[3] - , Gaussian measures on $L_p$ spaces $0 \leqslant p < \infty$ , ibid. 59 (1977), 249-261.

[4] - , Zero-one laws for Gaussian measures on metric abelian groups, ibid. 69.

[5] T. Byczkowski, T. Żak, Asymptotic properties of semigroups of measures on vector spaces, to appear in Ann. Prob.

[6] S. Csörgő, On weak convergence of the empirical process with random sample size, Acta Sci. Math. Szeged. 36 (1974), 17-25.

[7] P. Fernandez, A weak convergence theorem for random sums in normed space, Ann. Math. Stat. 42 (1971), 1737-1741.

[8] X. Fernique, Intégrabilité des vecteurs gaussuens, C.R. 270 (1970), 1698-1699.

[9] B.V. Gnedenko, H. Fahim, On a transfer theorem, DAN SSSR. 187 (1969), 15-17 (in russian) .

[10] M.G. Hahn, Central limit theorems in D[0,1] , Z.Wahr. 44 (1978), 89-101.

[11] T. Inglot, A. Weron, On Gaussian random elements in some non-Banach spaces, Bull. Acad. Sci. Ser. Math. Astronom. Phys. 22 (1974), 1039-1043.

[12] J. Kuelbs, The invariance principle for Banach space valued random variables, J. Mult. Anal. 3 (1973), 161-172.

[13] K.R. Parthasarathy, Probability measures on metric spaces, Academic Press, 1967.

[14] B.L.S. Prakasa Rao, Limit theorems for random number of random elements on complete separable metric spaces, Acta Math. Acad. Sci. Hung. 24 (1973), 1-4.

[15] R. Pyke The weak convergence of the empirical process of random sample size, Proc. Camb. Phil. Soc. 64 (1968), 155-160.

[16] W. Szczotka, On the central limit theorem for the breakdown processes, to appear in Probability and Math. Statistics.

INSTITUTE OF MATHEMATICS
TECHNICAL UNIVERSITY
50-370 WROCŁAW

# ITERATED LOGARITHM LAWS FOR THE SQUARE INTEGRAL OF A WIENER PROCESS

Endre Csáki

Mathematical Institute of Hung. Acad. Sci.
Reáltanoda u. 13-15.
BUDAPEST, HUNGARY
H-1053

## 1. Introduction

Consider a standard Wiener process $w(t),\ t\geq o,\ w(o)=o$ and its square integral

$$(1.1)\qquad I(T) = \int_o^T w^2(t)\,dt\ .$$

It is well known that Strassen's law of the iterated logarithm [5] implies that

$$(1.2)\qquad \limsup_{T\to\infty} I(T)(T^2 \operatorname{loglog} T)^{-1} = 8/\Pi^2 \qquad \text{a.s.}$$

On the other hand, Donsker and Varadhan [2] established functional law of the iterated logarithm for local times, which implies that

$$(1.3)\qquad \liminf_{T\to\infty} I(T)\, T^{-2} \operatorname{loglog} T = 1/8 \qquad \text{a.s.}$$

In this paper we extend the above results by investigating upper and lower classes for $I(T)$. Following Révész [4] we say that

$a_1(T) \in UUC$ (upper-upper class) if with probability 1, $I(T) \leq a_1(T)$ eventually,

$a_2(T) \in ULC$ (upper-lower class) if with probability 1, the inequality $I(T) \geq a_2(T)$ holds at least for a sequence of $T$ increasing to $\infty$,

$b_1(T) \in LUC$ (lower-upper class) if with probability 1, the inequality $I(T) \leq b_1(T)$ holds at least for a sequence of $T$ increasing to $\infty$,

$b_2(T) \in LLC$ (lower-lower class) if with probability 1, $I(T) \geq b_2(T)$ eventually.

It is easy to see that $I(T)$ has the same distribution as $T^2 I(1)$. The distribution of $I(1)$ has been determined by Cameron and Martin [1]:

$$P(I(1) < x) =$$

$$(1.4) \quad = 2\sqrt{2} \sum_{n=o}^{\infty} \binom{-1/2}{n} \left(1-\Phi\left(\frac{4n+1}{2\sqrt{x}}\right)\right) =$$

$$= 1 - \frac{2}{\Pi} \sum_{n=o}^{\infty} (-1)^n \int_{\Pi(2n+\frac{1}{2})}^{\Pi(2n+\frac{3}{2})} \frac{1}{u\sqrt{-\cos u}} e^{-\frac{xu^2}{2}} du.$$

From this exact formula we may obtain asymptotic values for the lower and upper tails of the distribution:

$$(1.5) \quad P(I(1)<x) \sim \frac{4\sqrt{x}}{\sqrt{\Pi}} e^{-\frac{1}{8x}} \qquad \text{as} \quad x \to o$$

$$(1.6) \quad P(I(1)>y) \sim \frac{4\sqrt{2}}{\Pi^2\sqrt{y}} e^{-\frac{\Pi^2 y}{8}} \qquad \text{as} \quad y \to \infty .$$

For real number $c$ put

(1.7) $$I_c(1) = \int_o^1 (w(t)+c)^2\, dt$$

By expanding the Wiener process in terms of eigenfunctions of its covariance operator it can be seen that the following representation holds:

(1.8) $$I_c(1) = \sum_{k=o}^{\infty} \frac{(x_k + c\sqrt{2})^2}{(k+\frac{1}{2})^2\, \Pi^2} ,$$

where $x_k$ are i.i.d. random variables having standard normal distribution. From (1.8) it is easy to show that the inequality

(1.9) $$P(I_c(1) \le x) \le P\,(I_o(1) \le x) \qquad \text{for} \qquad x \ge o$$

holds true.

## 2. Main results

Our results are complete for the lower part only.

__Theorem 1.__ __Assume that__ $b(T) \downarrow o$ __and__ $T^2 b(T) \uparrow \infty$ __as__ $T \to \infty$ __.Then__

$$T^2 b(T) \in LLC \qquad \text{or } LUC$$

__according as__

(2.1) $$\int^{\infty} \frac{1}{t\sqrt{b(t)}}\, e^{-\frac{1}{8b(t)}}\, dt < \infty \qquad \text{or} = \infty .$$

__Proof.__ Assume first that the integral in (2.1) converges. Define $T_1 = 1$ and

(2.2) $$T_{k+1} = (1+b(T_k))T_k \quad .$$

For brevity put $b_k = b(T_k)$. It suffices to show that

(2.3) $$\sum_k P(I(T_k) < T_{k+1}^2 b_{k+1}) < \infty .$$

From (1.5) we have

$$P(I(T_k) < T_{k+1}^2 b_{k+1}) = P\left(I(1) < \frac{T_{k+1}^2}{T_k^2} b_{k+1}\right) \sim$$

$$\sim \frac{4}{\sqrt{\Pi}} \frac{T_{k+1}}{T_k} \sqrt{b_{k+1}}\, e^{-\frac{T_k^2}{8T_{k+1}^2 b_{k+1}}} =$$

$$= \frac{4}{\sqrt{\Pi}} (1+b_k) \sqrt{b_{k+1}}\, e^{-\frac{1}{8b_{k+1}(1+b_k)^2}} \leq$$

$$\leq K \sqrt{b_{k+1}}\, e^{-\frac{1}{8b_{k+1}}},$$

since, by our assumptions, both $b_k$ and $b_k/b_{k+1}$ are bounded. On The other hand it can be seen that the integral in (2.1) and $\sum_k \sqrt{b_{k+1}}$ $\exp(-1/8\, b_{k+1})$ converge together, hence (2.3) follows, which in turn proves the first part of our Theorem 1.

Assume now that the integral in (2.1) diverges. Let $T_k$ be defined as before, then we show that

(2.4) $$P(I(T_k) < T_k^2 b_k \text{ i.o.}) = 1 .$$

By using (1.5), the divergence of the integral in (2.1) is easily seen to imply that

(2.5) $$\sum_k P(I(T_k) < T_k^2 b_k) = \infty .$$

To show (2.4) we apply the following version of the Borel-Cantelli lemma due to Erdős and Rényi [3]:

If $A_k$ are events such that $\sum_k P(A_k) = \infty$ and

(2.6) $$\liminf_{n\to\infty} \frac{\sum_{k=1}^{n}\sum_{\ell=1}^{n} P(A_k A_\ell)}{\sum_{k=1}^{n}\sum_{\ell=1}^{n} P(A_k)P(A_\ell)} = 1 ,$$

then $P(A_k \text{ i.o.}) = 1.$

We have to verify (2.6) for the events

(2.7) $$A_k = \{I(T_k) < T_k^2 b_k\} .$$

Let $k < \ell$, we show first that

(2.8) $$P(A_k A_\ell) \leq P(A_k)\ P(I(1) < \frac{T_\ell^2 b_\ell}{(T_\ell - T_k)^2}) .$$

By using the inequality (1.9) and the fact that $A_k$ and $\{\int_{T_k}^{T_\ell} w^2(t)\,dt < T_\ell^2 b_\ell\}$ are independent under the condition $w(T_k) = z$, we can proceed as follows:

$$P(A_k A_\ell) \leq P(A_k, \int_{T_k}^{T_\ell} w^2(t)\,dt < T_\ell^2 b_\ell) =$$

$$= \int_{-\infty}^{\infty} P(A_k, \int_{T_k}^{T_\ell} w^2(t)\,dt < T_\ell^2 b_\ell \ / \ w(T_k) = z)\ dP\ (w(T_k) < z) =$$

$$= \int_{-\infty}^{\infty} P(A_k/w(T_k)=z)\ P(\int_{T_k}^{T_\ell} w^2(t)\,dt < T_\ell^2 b_\ell/w(T_k)=z)\ dP\ (w(T_k) < z) \leq$$

$$\leq P(\int_{T_k}^{T_\ell} w^2(t)\,dt < T_\ell^2 b_\ell/w(T_k)=0)\ P(A_k) =$$

$$= P(A_k)\ P((T_\ell - T_k)^2\ I(1) < T_\ell^2 b_\ell)\ ,$$

i.e. (2.8) follows.

From (1.4) we have

$$(2.9) \qquad P((T_\ell - T_k)^2\ I(1) < T_\ell^2 b_\ell) \leq \frac{4}{\sqrt{\Pi}} \frac{T_\ell}{T_\ell - T_k} \sqrt{b_\ell}\ e^{-\frac{(T_\ell - T_k)^2}{8T_\ell^2 b_\ell}}$$

For further calculations assume that

$$(2.10) \qquad (10 \log k)^{-1} \leq b_k \leq (4 \log k)^{-1} .$$

We shall get rid from this assumption later. The following inequality will be used:

$$(2.11) \qquad \frac{T_k}{T_\ell} \leq (1+b_\ell)^{-(\ell-k)}$$

easily obtained from (2.2).

Let

$$(2.12) \qquad \alpha = 1 - \max_\ell\ (1+b_\ell)^{-\frac{1}{b_\ell}}$$

and for given $\varepsilon > o$, define $k_o$ such that for $k > k_o$ we have

$$k^{\alpha^2/2} > \log(k+k^{\alpha^2/2}+1)$$

$$\frac{4}{\sqrt{\Pi}} \sqrt{b_k} \; e^{-\frac{1}{8b_k}} \leq (1+\varepsilon)\; P(A_k)$$

$$(\log k)^2 \geq \frac{-\log(8b_k \log(1+\varepsilon))}{\log\,(1+b_k)}$$

$$\frac{1}{1-4\varepsilon b_k} \leq (1+\varepsilon)\;.$$

For given $k_o < k \leq n$ split the set $\{\ell\colon k < \ell \leq n\}$ into three parts:

$$L_1 = \{\ \ell\colon 1 \leq \ell-k < b_\ell^{-1}\ \}$$

$$L_2 = \{\ \ell\colon b_\ell^{-1} \leq \ell-k < k^{\alpha^2/2}\ \}$$

$$L_3 = \{\ \ell\colon k^{\alpha^2/2}+k \leq \ell \leq n\ \}\;.$$

If $\ell \in L_1$, then from (2.11),

$$\frac{1}{b_\ell}\left(1-\frac{T_k}{T_\ell}\right)^2 \geq \frac{1}{b_\ell}\left(1-\frac{1}{(1+b_\ell)^{\ell-k}}\right)^2 \geq \frac{b_\ell(\ell-k)^2}{4}\,, \tag{2.13}$$

where we used the inequality

$$(1+b_\ell)^{\ell-k}\left(1-\frac{b_\ell(\ell-k)}{2}\right) \geq 1 \qquad \text{for } b_\ell(\ell-k) \leq 1\;. \tag{2.14}$$

Hence we obtain after some calculations

$$(2.15)\qquad \sum_{\ell\in L_1} P(A_k A_\ell) \le K\, P(A_k)\,, \qquad k > k_o$$

with some constant $K$.

If $\ell \in L_2$, then

$$(2.16)\qquad 1-\frac{T_k}{T_\ell} \ge 1-(1+b_\ell)^{-(\ell-k)} \ge 1-(1+b_\ell)^{-\frac{1}{b_\ell}} \ge \alpha$$

and we get for $k > k_o$,

$$(2.17)\qquad \begin{aligned} \sum_{\ell\in L_2} P(A_k A_\ell) &\le P(A_k)\frac{4}{\sqrt{\Pi}} \sum_{\ell\in L_2} \frac{\sqrt{b_\ell}}{\alpha}\, e^{-\frac{\alpha^2}{8b_\ell}} \le \\ &\le K_1 P(A_k) \sum_{\ell\in L_2} \ell^{-\alpha^2/2} \le K_1 P(A_k) \end{aligned}$$

with some constant $K_1$.

Finally, if $\ell \in L_3$, and $k > k_o$, then

$$(2.18)\qquad (\ell-k) \ge \frac{-\log(8b_\ell \log(1+\varepsilon))}{\log(1+b_\ell)}\,,$$

since it can be seen that $\ell-k \ge (\log\ell)^2$.
Therefore

$$P(A_k A_\ell) \le \frac{P(A_k)}{1-4\varepsilon b_\ell}\left(\frac{4}{\sqrt{\Pi}}\sqrt{b_\ell}\, e^{-\frac{1}{8b_\ell}}\right)(1+\varepsilon) \le$$

(2.19)

$$\leq (1+\varepsilon)^3 \; P(A_k)P(A_\ell) \quad .$$

To verify (2.6) we proceed as follows:

$$\sum_{k=1}^{n} \sum_{\ell=1}^{n} P(A_k A_\ell) = \sum_{k=1}^{n} P(A_k)+2 \sum\sum_{1\leq k<\ell\leq n} P(A_k A_\ell) =$$

$$= \sum_{k=1}^{n} P(A_k)+2 \sum_{k=1}^{k_o} \sum_{\ell=k+1}^{n} P(A_k A_\ell) + 2 \sum_{k=k_o+1}^{n} \sum_{\ell\in L_1} P(A_k A_\ell) +$$

$$+ 2 \sum_{k=k_o+1} \sum_{\ell\in L_2} P(A_k A_\ell)+2 \sum_{k=k_o+1}^{n} \sum_{\ell\in L_3} P(A_k A_\ell) \leq$$

$$\leq (1+2k_o+2K+2K_1) \sum_{k=1}^{n} P(A_k)+(1+\varepsilon)^3 \sum_{k=1}^{n} \sum_{\ell=1}^{n} P(A_k)\; P(A_\ell)$$

from which (2.6) follows easily.

Hence we proved $P(A_k \text{ i.o.})=1$, provided $b_k$ satisfies (2.10). In the case when (2.10) is not satisfied, define $b'_k=\min(b_k,(4\log k)^{-1})$ and omit those $A_k$ for which $b_k \leq (10\log k)^{-1}$. Then, as easily seen we may apply the above procedure to the new sequence $A_k^{(1)}$, i.e. prove that $\sum_k P(A_k^{(1)})=\infty$ and that (2.6) holds for $A_k^{(1)}$. Hence $P(A_k^{(1)}\text{ i.o.})=$ which inturn implies that $P(A_k \text{ i.o.})=1$. This completes the proof of Theorem 1.

Our second theorem concerns *UUC* and *ULC*, but is not complete.

<u>Theorem 2.</u>

(i) <u>Assume that</u> $a(T)\uparrow\infty$ <u>as</u> $T\to\infty$. <u>If</u>

(2.20) $$\int \frac{\sqrt{a(t)}}{t} e^{-\frac{\Pi^2 a(t)}{8}} dt < \infty ,$$

then

$$T^2 a(T) \in UUC.$$

(ii)

(2.21) $$\frac{8T^2}{\Pi^2} (\log\log T + \tfrac{1}{2}\log\log\log\log T) \in ULC .$$

Since the proof of (i) is the same as the proof of the first part of Theorem 1, we omit it.

To prove (ii), we apply the following version of Borel-Cantelli lemma due to Donsker and Varadhan [2]:

Let $F_n$ be an inreasing sequence of $\sigma$ - fields and $A_n \in F_n$ . If $\sum_n P(A_n/F_{n-1}) = \infty$ a.s., then $P(A_n$ i.o.$) = 1$.

Now define $T_1=1$, $T_n=T_{n-1}(1+\log n)$, $n=2,3,\ldots$ Let the events $A_n$ be defined by

(2.22) $$A_n = \{I(T_n)-I(T_{n-1}) > T_n^2\, a(T_n)\} ,$$

where

(2.23) $$a(T) = \frac{8}{\Pi^2} (\log\log T + \tfrac{1}{2}\log\log\log\log T) .$$

We show that $P(A_n$ i.o.$) = 1$ which obviously implies that

(2.24) $$P(I(T_n) > T_n^2\, a(T_n) \text{ i.o.}) = 1 ,$$

i.e. part (ii) of Theorem 2.

Let $F_n$ be the $\sigma$ - field generated by $W(t)$, $t \leq T_n$. Then $A_n \in F_n$ , and by (1.6) and (1.9), we have

$$P(A_n/F_{n-1}) = P(A_n/W(T_{n-1})) \geq P(A_n/W(T_{n-1}) = 0) =$$

$$= P((T_n - T_{n-1})^2 I(1) > T_n^2 a(T_n)) \sim$$

$$\sim \frac{4\sqrt{2}}{\Pi^2\sqrt{a(T_n)}} \exp\left(-\frac{\Pi^2 T_n^2 a(T_n)}{8(T_n - T_{n-1})^2}\right) .$$

It is easy to see that

$$a(T_n) \sim \frac{8}{\Pi^2} \log n ,$$

$$\exp\left(-\frac{\Pi^2 T_n^2 a(T_n)}{8(T_n - T_{n-1})^2}\right) \sim \frac{K}{\log T_n \sqrt{\log\log T_n}}$$

$$\sim \frac{K}{n \log\log n \sqrt{\log n}} ,$$

i.e. $\sum_n P(A_n/F_{n-1}) = \infty$ a.s., which by the quoted Borel-Cantelli lemma implies that $P(A_n \text{ i.o.}) = 1$ , hence (ii) follows.

It is easy to see that Theorem 1 implies (1.3) and Theorem 2 implies (1.2). There is a gap however between (i) and (ii) of Theorem 2 even in the second term, because the integral in (2.20) converges for $a(T) = \frac{8}{\Pi^2}(\log\log T + c\log\log\log T)$ if and only if $c > \frac{3}{2}$ .

<u>CONJECTURE.</u> If the integral in (2.20) diverges, then $T^2 a(T) \in ULC$.

## REFERENCES

[1] CAMERON, R.H. and MARTIN, W.T., The Wiener measure of Hilbert neighborhoods in the space of real continuous functions, J.Math. Phys. 23(1944), 195-209.

[2] DONSKER, M.D. and VARADHAN, S.R.S., On laws of the iterated logarithm for local times, Comm.Pure Appl. Math. 30(1977), 707-753.

[3] ERDŐS, P. and RÉNYI, A., On Cantor's series with convergent $\sum 1/q_n$ , Ann. Univ. Sci. Budapest, R. Eötvös nom. Sect. Math. 2(1959), 93-109.

[4] RÉVÉSZ, P., A note to the Chung - Erdős - Sirao theorem, to appear.

[5] STRASSEN, V., An invariance principle for the law of the iterated logarithm, Z. Wahrscheinlichkeitstheorie verw. Gebiete 3(1964), 211-226.

# Asymptotic properties of the nonparametric survival curve estimators under variable censoring

Antónia Földes
Lidia Rejtő

SUMMARY

Let $X_1, Y_1, X_2, Y_2, \ldots$ be an independent sequence of random variables, the $X_n$'s having common continuous survival function $F$ and the $Y_n$'s having continuous survival functions $G_n$'s /survival function = 1-distribution function/.

It is shown that the product limit estimator of $F$ from data $\{Z_n, \delta_n\}_{n=1}^{\infty}$ where $Z_n = \min(X_n | Y_n)$ and $\delta_n = [X_n \leq Y_n]$ /[ ] denotes the indicator function/, /see Kaplan and Meier, J.A.S.A. 53 /1958/ 457-481/ is strong uniformly consistent on an interval $(-\infty, T]$ under some reasonable conditions.

Assuming that $F$ is distributed according to a Dirichlet process with parameter $\alpha$ /see Ferguson, Ann. Statist. 1 /1973/ 209-230/ it is shown that the sup distance between the Bayesian estimator /see Susarla, Van Ryzin J.A.S.A. 72 /1976/ 889-902/ and the product limit estimator is small enough to remain valid all results using the Bayesian estimator.

AMS 1970 subject classification
Primary 60F15 Secondary 62G05

## 1. Introduction

Let $X_1, X_2, \ldots, X_n, \ldots$ be an i.i.d. sequence of random variables with distribution function $1-F(t)$ . Let $Y_1, Y_2,$ $\ldots, Y_n, \ldots$ be another sequence of independent random variables, with distributions $1-G_1(t), 1-G_2(t), \ldots, 1-G_n(t), \ldots$ . Suppose that the sequences $\{X_i\}_{i=1}^n$ and $\{Y_i\}_{i=1}^n$ are mutually independent of each other. Set

$$\delta_i = \begin{cases} 1 & \text{if} \quad X_i \le Y_i \\ 0 & \text{if} \quad X_i > Y_i \end{cases} \qquad i = 1, 2, \ldots, n$$

and

$$Z_i = \min\{X_i, Y_i\}.$$

An important problem in reliability and survival analysis is the estimation of the distribution function of $X$ by the censored sample $\{Z_i, \delta_i\}_{i=1}^n$ .

The maximum likelihood estimator of $F$ in case of i.i.d. , was obtained by Kaplan and Meier [8], this is the so-called product-limit estimator. In the noncensored case /i.e. all $\delta_i = 0$, $i = 1, \ldots, n$ / the $PL$ is equal to the usual empirical distribution function of $F$ . The aim of this paper is to show that the properties of the $PL$ estimator are similar to those of the empirical distribution function. In case of i.i.d. $Y$-s , this problem was already investigated in [15] and [7]. Recently in [6] it was proved that the $PL$ estimator is uniformly almost surely consistent with rate $O\left(\sqrt{\frac{\log n}{n}}\right)$ for continuous $F$ and $G$ . This paper concerns the case of variable censoring. The notations and definitions are given in Section 2. Section 3 contains the main theorems. Theorem 3.1 gives exponential bound of the probability $P(\sup|F_n^* - F| > \varepsilon)$ in case of variable $Y$-s . Theorem 3.2 gives a general strong law result to $\sup|F_n^*(u) - F(u)|$ on an interval $(-\infty, T)$ where $T < +\infty$ . Corollary 3.2.2 contains the $O\left(\sqrt{\frac{\log n}{n}}\right)$ result to the i.i.d. censoring case. Section 4 contains the necessary lemmas and the proofs of the theorems are postponed to Section 5.

Section 6 is devoted to the Bayesian estimator of $F$ . The properties of this estimator were investigated in [12], [13], [14]. In [ 9 ] Phadia and Van Ryzin proved that the distance between the PL and the Bayesian estimator is small in each fixed point $n$ . In Lemma 6.1 it is proved that the sup distance of the two estimators is $O(\frac{1}{n})$ . Therefore the validity of Theorem 3.3 and all its corollaries extends to the case of the Bayesian estimator.

## 2. Definitions and Notations

Let $X_1, \ldots, X_n, \ldots$ be independent identically distributed /i.i.d./ random variables. Let $Y_1, Y_2, \ldots, Y_n, \ldots$ be another sequence of independent /not necessarily identically distributed/ random variables such that $\{X_i\}_{i=1}^{\infty}$ and $\{Y_i\}_{i=1}^{\infty}$ are mutually independent of each other.

Set

$$\delta_i = [X_i \le Y_i] \qquad \text{and} \qquad Z_i = \min\{X_i, Y_i\}$$

for $i = 1, 2, \ldots, n, \ldots$ , where $[A]$ denotes the indicator function of the set $A$ . The estimators of the distribution function $F$ are based on the sample $(\delta_1, Z_1), \ldots, (\delta_n, Z_n)$. For sake of simplicity we shall denote by $F, G_j, H_j$ the survival functions and not the distribution functions.

Set

$$P(X_j > u) = F(u) \qquad j = 1, \ldots, n, \ldots$$

$$P(Y_j > u) = G_j(u) \qquad j = 1, \ldots, n, \ldots$$

$$P(Z_j > u) = H_j(u) \qquad j = 1, \ldots, n, \ldots$$

Clearly

$$H_j(u) = F(u) G_j(u) \qquad j = 1, \ldots, n, \ldots$$

Let us denote by $N^+(u,n) = N^+(u) = \# Z_j$ 's greater than $u$. In case of i.i.d. $Y_j$ -s Kaplan and Meier obtained the so called product limit estimator of the distribution of $X$.

Definition. The product limit (PL) estimator $F_n^*$ of $F$ is

$$F_n^*(u) = \begin{cases} \prod_{j=1}^{n} \left( \dfrac{N^+(Z_j)}{N^+(Z_j)+1} \right)^{[\delta_j = 1, Z_j \leq u]} & \text{if } u \leq \max\{Z_j \ldots Z_n\} \\ 0 & \text{if } u > \max\{Z_1 \ldots Z_n\} . \end{cases}$$

This estimator is used in case of variable censoring as well.

If there is no censoring then $\delta_j = 1$ $j = 1, 2, \ldots$ and $1 - F_n^*(u)$ is equal to the empirical distribution function.

In course of the proofs we need the modified product limit estimator.

Definition. The modified product limit estimator of $F$ is

$$\overline{F}_n(u) = \begin{cases} \prod_{j=1}^{n} \left( \dfrac{N^+(Z_j)+1}{N^+(Z_j)+2} \right)^{[\delta_j = 1, Z_j \leq u]} & \text{if } u \leq \max\{Z_1, \ldots, Z_n\} \\ 0 & \text{if } u > \max\{Z_1, \ldots, Z_n\} \end{cases}$$

Set

$$T_F = \sup\{u: \ F(u) > 0\}$$
$$T_{G_j} = \sup\{u: \ G_j(u) > 0\} \qquad j = 1, \dots, n, \dots$$
$$T_{H_j} = \sup\{u: \ H_j(u) > 0\} \qquad j = 1, \dots, n, \dots$$
$$\beta_j(=) = [\delta_j = 1, \ Z_j \leq u] \qquad j = 1, \dots, n, \dots$$
$$\mu_n(u) = \sum_{j=1}^{n} H_j(u) \ , \quad \sigma_n^2(u) = \sum_{j=1}^{n} H_j(u)\,(1 - H_j(u))$$
$$\sigma_n(T) = \sigma_n \qquad \mu_n(T) = \mu_n$$

## 3. Principle results

The results of this section concern the behavior of $\sup |F_n^*(u) - F(u)|$. Necessary condition of the statements is that $\sigma_n \to \infty$ hold. At the same time it is not necessary to take strong assumptions about the rate of convergence. Therefore some of the theorems have conditions the fulfillment of which may be difficult to verify. The remarks and corollaries deal with special cases and make the conditions verifiable. The proofs of theorems are postponed to Section 5.

Theorem 3.1. Suppose that

/i/ the distribution functions $1-F, 1-G_1, \dots, 1-G_n$ of $X, Y_1, \dots, Y_n$ are continuous on $(-\infty, T]$.

/ii/ if $n > n_0$ then $\mu_n > 0$.

/iii/ $\{\alpha_n\}$ is an /arbitrary/ sequence of nonnegative numbers such that $0 \leq \alpha_n \leq \sigma_n$ for $n > n_0$.

/iv/ $n > n_0$ and $\varepsilon$ are such that

$$\frac{4 n^{3/2}}{\lambda \mu_n (\mu_n - \alpha_n \sigma_n)} < \varepsilon$$

where $\lambda > 0$ is a constant for which $\lambda^2 \le \sum_{k=1}^{n} H_k^2(T)$

Then for $n > n_0$

$$P\Big(\sup_{-\infty<u\le T} |F_n^*(u)-F(u)| > \varepsilon\Big) \le 6\exp\Big\{-\frac{2}{9}\alpha_n^2\Big\} + \frac{12}{\varepsilon}\exp\Big\{-\frac{2}{9}\frac{\varepsilon^2\mu_n^2}{n}\Big\} +$$

/3.1/

$$+\, 2e^2\,\frac{\varepsilon\mu_n(\mu_n-\alpha_n\sigma_n)}{n}\,\exp\Big\{-\frac{\varepsilon^2\mu_n(\mu_n-\alpha_n\sigma_n)^2}{2n^3}\Big\}.$$

Remark 3.1.1. Let us suppose that there exists a constant $a<1$, for which $na \le \mu_n$ /e.g. in case of i.i.d. $Y_j$'s/. Choose $\alpha_n = \sigma_n$. By Cauchy inequality:

$$\mu_n - \sigma_n^2 = \sum_{k=1}^{n} H_k^2(T) \ge \frac{1}{n}\Big(\sum_{h=1}^{n} H_h(T)\Big)^2$$

hence

/3.2/ $$\mu_n - \sigma_n^2 \ge n a^2$$

The condition /iv/ means that $\frac{1}{a^4\sqrt{n}} < \varepsilon$. Therefore in this case the result is

/3.3/ $$P\Big(\sup_{-\infty<u\le T} |F_n^*(u)-F(u)| > \varepsilon\Big) \le 6\exp\Big\{-\frac{2}{9}\sigma_n^2\Big\} +$$

$$+\, \frac{12}{\varepsilon}\exp\Big\{-\frac{2}{9}\varepsilon^2 n a^2\Big\} + 2e^2 n\varepsilon\exp\Big\{-\frac{1}{2}n\varepsilon^2 a^6\Big\}.$$

Remark 3.1.2. In case of i.i.d. $Y$'s, supposing that $\varepsilon > \frac{4}{\delta^4\sqrt{n}}$ it follows that

$$P\Big(\sup_{-\infty<u\le T} |F_n^*(u)-F(u)| > \varepsilon\Big) \le \frac{6}{\varepsilon}\exp\Big\{-\frac{2}{9}n\varepsilon^2\delta^6\Big\}.$$

This result is weaker than the one proved in [16] where instead of the $\frac{6}{\varepsilon}$ factor there is a constant only.

Theorem 3.2. Suppose that

/i/ the distribution functions $1-F, 1-G_1, \dots, 1-G_n$ of $X, Y_1, \dots, Y_n$ are continuous on $(-\infty, T]$

/ii/ $$\frac{n^{3/4}(\log n)^{1/4}}{\mu_n} = o(1)$$

/iii/ $\{\alpha_n\}$ is a sequence of nonnegative numbers for which $0 \le \alpha_n \le \sigma_n$ and $\sum_{n=1}^{\infty} \exp\{-\frac{2}{9}\alpha_n^2\} < +\infty$.

Then

/3.4/ $$\sup_{-\infty < u \le T} |F_n^*(u) - F(u)| = O\left(\frac{n^{3/2}\sqrt{\log n}}{\mu_n(\mu_n - \alpha_n\sigma_n)}\right). \qquad \text{a.s.}$$

__Corollary__ 3.2.1. Suppose that the conditions /i/ and /iii/ with $\alpha_n = \sigma_n$ of Theorem 3.2 are fulfilled, further suppose that there exists a constant $a$ such that $na \le \mu_n$ for all $n > n_0$. Then

$$\sup_{-\infty < u \le T} |F_n^*(u) - F(u)| = O\left(\sqrt{\frac{\log n}{n}}\right) \qquad \text{a.s.}$$

__Proof__. Choosing $\alpha_n = \sigma_n$, by Remark 3.1.1

$$\mu_n - \sigma_n^2 \ge na^2.$$

Therefore $$\frac{n^{3/2}\sqrt{\log n}}{\mu_n(\mu_n - \alpha_n\sigma_n)} \le \frac{\sqrt{\log n}}{a^3\sqrt{n}}.$$

__Corollary__ 3.2.2. Suppose that

/i/ $Y_1, \dots, Y_n$ have the same continuous distribution

/ii/ at the point $T$ $\min\{H(T); 1-H(T)\} > \delta > 0$.

Then $$\sup_{-\infty < u \le T} |F_n^*(u) - F(u)| = O\left(\sqrt{\frac{\log n}{n}}\right) \qquad \text{a.s.}$$

__Corollary__ 3.2.3. Suppose that there exists a sequence $\{\alpha_n\}$ such that the conditions /i/-/iii/ of Theorem 3.2 are fulfilled. Further suppose that there exists a constant $0 < b < 1$ such that

/3.5/ $$b\mu_n^{\beta} \le \mu_n - \alpha_n\sigma_n \qquad \text{if} \quad n > n_0$$

where $0 < \beta \le 1$ is fixed.

Then

/3.6/ $$\sup |F_n^*(u) - F(u)| = O\left(\frac{n^{3/2}\sqrt{\log n}}{\mu_n^{1+\beta}}\right) \qquad \text{a.s.}$$

Remark 3.2.1. From Theorem 3.2 the consistency of the PL doesn't follow if $\mu_n(T) = \sum_{k=1}^{n} H_k(T)$ goes to infinity slowly, namely if $\mu_n < o(n^{3/4})$. If in the interval $(-\infty;T]$ there exists a point $u$ for which $\mu_n(u) = \sum_{j=1}^{n} P(Z_j > u) = O(1)$ then after the point $u$ there are at most finitely many sample elements with probability 1, hence $F(u)$ cannot be estimated after $u$. Therefore a necessary condition of the uniform consistency of $F_n^*$ in $(-\infty, T]$ is the divergence of $\mu_n(T)$.

Remark 3.2.2. Let us suppose that $\mu_n \geq c n^{\alpha}$ and $\sum \exp\{-\sigma_n^2\} < +\infty$. Choosing $\alpha_n = \sigma_n$, it follows that $\mu_n - \sigma_n^2 = \sum_{k=1}^{n} H_n^2(T) \geq \mu_n^2/n$,

$$\sup_{-\infty < u \leq T} |F_n^*(u) - F(u)| = O\left(\frac{n^{5/2}\sqrt{\log n}}{n^{3\alpha}}\right). \qquad \text{a.s.}$$

Consequently Theorem 3.2. gives strong consistency for $\alpha > \frac{5}{6}$.

Moreover, let us suppose that $\sum_{n=1}^{\infty} \exp\left\{-\frac{2}{9}\sigma_n^{2\varrho}\right\} < +\infty$, for some $1 > \varrho > 0$. Choosing $\alpha_n = \sigma_n^{\varrho}$ we get that

$$\mu_n - \alpha_n \sigma_n = \mu_n - \sigma_n^{1+\varrho} \geq \mu_n - (\mu_n - \mu_n^2/n)^{\frac{1+\varrho}{2}} \geq \mu_n/2 \qquad \text{for } n > n_0(\varrho).$$

Therefore

$$\sup_{-\infty < u \leq T} |F_n^*(u) - F(u)| = O\left(\frac{n^{3/2}\sqrt{\log n}}{\mu_n^2}\right) \qquad \text{a.s.}$$

In this case the strong consistency holds if $\mu_n \geq cn^{\alpha}$ for $\alpha > 3/4$.

On the other hand supposing that the sequence $\{H_k(T)\}$ is monoton decreasing and $\mu_n = \Theta(n)$ we get that $\sigma_n \geq \sqrt{\mu_n/2}$ thus $\mu_n - \sigma_n^2 \geq 3\mu_n/4$. Consequently the series $\sum \exp\{-\frac{2}{9}\sigma_n^2\}$ is a convergent one if $\mu_n > cn^{\alpha}$ $(\alpha > 0)$. Further the condition /3.5/ of Corollary 3.2.3 fulfils with $\beta = 1$ therefore

$$\sup |F_n^*(u) - F(u)| = O\left(\frac{n^{3/2}\sqrt{\log n}}{\mu_n^2}\right) \qquad \text{a.s.}$$

It gives strong consistency again if $\alpha > 3/4$.

Remark 3.2.3. Set

$$T_F = \sup\{x : F(x) > 0\}.$$

$T_G$ resp. $T_{G_1}, T_{G_2}, \dots$ are defined similarly. In case of i.i.d. censoring it is possible to estimate $F$ within the interval $(-\infty; K]$ only where $K \le \min\{T_F; T_G\}$. In case of different censoring it is considered an interval $(-\infty; T]$ where $T < T_F$, but some of the $T_{G_k}$ /maybe infinitely many/ can be less than $T$. In the case when $T_{G_k} < T$ the $k$-th sample element shall be in $(-\infty, T_{G_k}]$. Nevertheless, if the behavior of the series $\sum_1^\infty H_j(T)$ and $\sum H_j(T)(1-H_j(T))$ is "good enough" Theorem 3.2. gives a convergence rate.

Remark 3.2.4. In case of i.i.d. $Y$-s, Corollary 3.2.2 proved to be weaker in comparison with Theorem 1 of [1]. In that paper a $O\left(\sqrt{\frac{\log\log n}{n}}\right)$ rate is proved on the interval $(-\infty, T_F)$ supposing that $G(T_F) > 0$.

## 4. Lemmas

The following decomposition of $\log \bar{F}_n(u)$ will be used /similar decomposition is used by Susarla and Van Ryzin [13]/

$$\log \bar{F}_n(u) = R_{n,1}(u) + R_{n,2}(u) + R_{n,3}(u) \tag{4.1}$$

where

$$R_{n,1}(u) = -\sum_{j=1}^{n} \frac{\beta_j(u)}{\sum_{k=1}^{n} H_k(Z_j)}, \tag{4.2}$$

$$R_{n,2}(u) = -\sum_{j=1}^{n} \sum_{\ell=2}^{\infty} \beta_j(u) \frac{1}{\ell} (2 + N^+(Z_j))^{-\ell}, \tag{4.3}$$

$$/4.4/ \quad R_{n,3}(u) = -\sum_{j=1}^{n} \beta_j(u) \left\{ \frac{1}{2+N^+(Z_j)} - \frac{1}{\sum_{k=1}^{n} H_k(Z_j)} \right\}$$

If $N^+(Z_j)=0$ then the logarithmic expansion of $\log F_n^*(u)$ is not possible, but it can be done for $\log \bar{F}_n(u)$.

Lemma 4.1 contains results which are essential for further arguments. In Lemma 4.2 it is proved that the sup distance $|\bar{F}_n(u) - F_n^*(u)|$ is small. Lemmas 4.3-4.5 deal with the behavior of $\sup|\log F(u) - R_{n,1}(u)|$, $\sup|R_{n,2}(u)|$ and $\sup|R_{n,3}(u)|$. Using that the inequality $|x-y| \le |\log x - \log y|$ holds if $0<x\le 1$, $0<y\le 1$ it is obvious that the behavior of $\sup|\bar{F}_n(u) - F(u)|$ can be reduced to the behavior of

$$\sup|\log \bar{F}_n(u) - \log F(u)| \le \sup|\log F(u) - R_{n,1}(u)| + \sup|R_{n,2}(u)| + \sup|R_{n,3}(u)|.$$

Lemma 4.3 implies that $E(R_{n,1}(u)) = \log F(u)$ thus it is the mean term in the decomposition /4.1/. In Lemmas 4.4 and 4.5 it is proved that the remaining terms $R_{n,2}(u)$ and $R_{n,3}(u)$ of the decomposition are uniformly small.

In all of the lemmas the sup is considered on an interval $(-\infty;T]$ and it is supposed that at the point $T$; $\sum_{k=1}^{n} H_k(T) > 0$ if $n > n_0$.

Lemma 4.1. Let $0 < \alpha_n \le \sigma_n$ be arbitrary.

/i/ If $a_n, \beta$ and $\varepsilon$ are positive numbers such that

$$0 < \frac{a_n}{(\mu_n - \alpha_n \sigma_n)^\beta} < \varepsilon \quad \text{then}$$

$$P\left(\frac{a_n}{(N^+(T)+1)^\beta} > \varepsilon\right) \le 2\exp\left\{-\frac{2}{9}\alpha_n^2\right\}.$$

/ii/ If $\sum_{n=1}^{\infty} \exp\{-\frac{2}{9}\alpha_n^2\} < +\infty$ then

$$\frac{1}{N^+(T)+1} = O\left(\frac{1}{\mu_n - \alpha_n \sigma_n}\right) \qquad \text{a.s.}$$

Proof. /i/ Let

/4.5/ $$B_n = \{\omega;\ |N^+(T) - \mu_n| > \alpha_n \sigma_n\}.$$

From the Bernstein inequality /see e.g. [10]/ it follows that

/4.6/ $$P(B_n) \le 2\exp\{-\tfrac{2}{9}\alpha_n^2\}.$$

Now

/4.7/ $$P\left(\frac{a_n}{(N^+(T)+1)^\beta} > \varepsilon\right) \le P\left(\frac{a_n}{(N^+(T)+1)^\beta} > \varepsilon, \bar{B}_n\right) + P(B_n).$$

On the set $\bar{B}_n$

/4.8/ $$\mu_n - \alpha_n \sigma_n \le N^+(T) \le \mu_n + \alpha_n \sigma_n .$$

From /4.7/ and /4.8/ using the condition of part /i/

$$P\left(\frac{a_n}{(N^+(T)+1)^\beta} > \varepsilon, \bar{B}_n\right) \le P\left(\frac{a_n}{(\mu_n - \alpha_n \sigma_n)^\beta} > \varepsilon\right) = 0$$

follows, and the statement comes from /4.6/, /4.7/ and /4.8/.

/ii/ Using part /i/ of the Lemma with $\beta = 1$, $\varepsilon = 2$, and $\alpha_n = \mu_n - \alpha_n \sigma_n$ we get that

$$\sum_{n=1}^{\infty} P\left(\frac{\mu_n - \alpha_n \sigma_n}{N^+(T)+1} > 2\right) \le \sum_{n=1}^{\infty} 2\exp\left(-\frac{2}{9}\alpha_n^2\right)$$

By assumption the last sum converges and the assertion follows from the Borel-Cantelli lemma.

Lemma 4.2. Let $0 < \alpha_n \le \sigma_n$.

/i/ If $n$ and $\varepsilon$ are such that the inequality $0 < \frac{n}{(\mu_n - \alpha_n \sigma_n)^2} < \varepsilon$ holds then

$$P\left|\sup_{-\infty < u \le T} |F_n^*(u) - \bar{F}_n(u)| \le 2\exp\{-\tfrac{2}{9}\alpha_n^2\}.\right.$$

/ii/ If $\sum_{n=1}^{\infty} \exp\{-\frac{2}{9}\alpha_n^2\} < \infty$ then

$$P\Big(\sup_{-\infty<u\le T}|F_n^*(u)-\bar F_n(u)| = O\Big(\frac{n}{(\mu_n-\alpha_n\sigma_n)^2}\Big) \qquad \text{a.s.}$$

Proof.

$$|F_n^*(u)-\bar F_n(u)| = \Big|\prod_{j=1}^n\Big(\frac{N^+(Z_j)}{N^+(Z_j)+1}\Big)^{\beta_j(u)} - \prod_{j=1}^n\Big(\frac{N^+(Z_j)+1}{N^+(Z_j)+2}\Big)^{\beta_j(u)}\Big|$$

Using that for $0\le a_i\le 1$, $0\le b_i\le 1$, $i=1,\dots,n$,

$$/4.9/\qquad \Big|\prod_{i=1}^n a_i - \prod_{i=1}^n b_i\Big| \le \sum_{i=1}^n |a_i-b_i|$$

by an easy computation we get that $\sup_{-\infty<u\le T}|F_n^*(u)-\bar F_n(u)|\le$ $\le \frac{n}{(N^+(T)+1)^2}$. Therefore the statements follows from Lemma 4.1.

Lemma 4.3

/i/ $P\big(\sup_{-\infty<u\le T}|R_{n1}(u)-\log F(u)|>\varepsilon\big)\le \frac{24|\log F(T)|}{\varepsilon}\exp\Big(-\frac{2}{9}\frac{\varepsilon^2\mu_n^2}{n}\Big)$

/ii/ If $\frac{\sqrt{n\log n}}{\mu_n} = o(1)$ then

$$\sup_{-\infty<u\le T}|R_{n1}(u)-\log F(u)| = O\Big(\frac{\sqrt{n\log n}}{\mu_n}\Big) \qquad \text{a.s.}$$

Proof. The argument is similar to the one in [6]. First it will be shown that $E(R_{n1}(u)) = \log F(u)$

$$E(R_{n1}(u)) = \sum_{j=1}^n -E\Big(\frac{[\delta_j=1, Z_j\le u]}{\sum_{k=1}^n H_k(Z_j)}\Big) = -\sum_{j=1}^n\int_{-\infty}^u \frac{G_j(t)}{\sum_{k=1}^n H_k(t)}\,d(1-F(t)) =$$

$$/4.10/\qquad = \int_{-\infty}^u \frac{\sum_{j=1}^n G_j(t)}{F(t)\sum_{k=1}^n G_j(t)}\,d(1-F(t)) = \log F(u).$$

Observe that for every fixed $u\in(-\infty,T]$ the summands of $R_{n1}(u)$ are bounded, that is $0\le\Big|\frac{\beta_j(u)}{\sum_{k=1}^n H_k(Z_j)}\Big|\le\frac{1}{\sum_{k=1}^n H_k(T)}$.

Hence Theorem 2 of paper [1] can be applied and

$$/4.11/\qquad P(|R_{n1}(u)-\log F(u)|>\varepsilon)\le 2\exp\Big(\frac{-2\varepsilon^2(\sum_{k=1}^n H_k(T))^2}{n}\Big) = 2\exp\Big(\frac{-2\varepsilon^2\mu_n^2}{n}\Big)$$

follows. Observe that $R_{n1}(u)$ and $\log F(u)$ are both monoton decreasing functions of $u$.

Let us choose a partition $-\infty = \eta_0 < \eta_1 < \dots < \eta_{L(\varepsilon)} = T$ in such a way that

a/ $\log F(\eta_i) - \log F(\eta_{i+1}) < \frac{\varepsilon}{3}$ $\quad i = 0,1,\dots L(\varepsilon)-1.$

b/ $L(\varepsilon) \le \frac{6|\log F(T)|}{\varepsilon}$

If $|R_{n1}(\eta_i) - \log F(\eta_i)| < \frac{\varepsilon}{3}$ and $|R_{n1}(\eta_{i+1}) - \log F(\eta_{i+1})| < \frac{\varepsilon}{3}$ and $\eta_{i-1} < u \le \eta_i$ then $|R_{n1}(u) - \log F(u)| < \frac{\varepsilon}{3} + 2\frac{\varepsilon}{3} = \varepsilon$.

Hence if $\sup_{-\infty<u\le T} |R_{n1}(u) - \log F(u)| > \varepsilon$ then for some $0 < i \le L(\varepsilon)$ $|R_{n1}(\eta_i) - \log F(\eta_i)| \ge \frac{\varepsilon}{3}$. Therefore applying /4.11/

$$P(\sup_{-\infty<u\le T} |R_{n1}(u) - \log F(u)| > \varepsilon) \le 2L(\varepsilon) \cdot 2\exp\left(-\frac{2}{9}\frac{\varepsilon^2\mu_n^2}{n}\right) \le$$

$$\le \frac{24|\log F(T)|}{\varepsilon} \exp\left(-\frac{2}{9}\frac{\varepsilon^2\mu_n^2}{n}\right)$$

which proves /i/. /ii/ follows by the standard Borel-Cantelli argument.

Lemma 4.4. Let $0 \le \alpha_n \le \sigma_n$ be arbitrary.

/i/ If $n$ and $\varepsilon$ are such that the inequality $0 < \frac{n}{(\mu_n - \alpha_n\sigma_n)^2} < \varepsilon$ holds then $P\left(\sup_{-\infty<u\le T} |R_{n2}(u)| > \varepsilon\right) \le 2\exp\left(-\frac{2}{9}\alpha_n^2\right)$.

/ii/ If $\sum_{n=1}^{\infty} \exp\left(-\frac{2}{9}\alpha_n^2\right) < \infty$ then

$$\sup_{-\infty<u\le T} |R_{n2}(u)| = O\left(\frac{n}{(\mu_n - \alpha_n\sigma_n)^2}\right) \quad \text{a.s.}$$

Proof. The results follow from the obvious inequality

$$\sup_{-\infty<u\le T} |R_{n2}(u)| \le \frac{n}{(N^+(T)+1)^2}$$

and from Lemma 4.1.

Lemma 4.5. Let $0 < \alpha_n \le \sigma_n$

/i/ If $n$ and $\varepsilon$ are such that the inequality $0 < \frac{4n^{3/2}}{\mu_n(\mu_n - \alpha_n\sigma_n)} < \varepsilon$ holds then $P(\sup_{-\infty<u\le T} |R_{n3}(u)| > \varepsilon) \le$

$$\le 2\exp\left(-\frac{2}{9}\alpha_n^2\right) + 2e^2 \frac{\varepsilon\mu_n(\mu_n - \alpha_n\sigma_n)}{n} \exp\left(-\frac{\varepsilon^2\mu_n^2(\mu_n - \alpha_n\sigma_n)^2}{2n^3}\right).$$

/ii/ If $\sum_{n=1}^{\infty} \exp\left\{-\frac{2}{9}\alpha_n^2\right\} < +\infty$ then

$$\sup_{-\infty<u\le T} |R_{n3}(u)| = O\left(\frac{n^{3/2}\sqrt{\log n}}{\mu_n(\mu_n - \alpha_n\sigma_n)}\right) \quad \text{a.s.}$$

Proof. /i/ By a simple manipulation it can be seen that

$$/4.12/ \quad \sup_{-\infty<u\le T} |R_{n3}(u)| \le n \sup_{-\infty<u\le T} \left| \frac{1}{2+N^+(u)} - \frac{1}{\sum_{k=1}^{n} H_k(u)} \right|.$$

Now using the set $B_n$ defined by /4.5/ and inequalities /4.6/, /4.8/, /4.12/ one gets

$$P(\sup_{-\infty<u\le T}|R_{n3}(u)|>\varepsilon)\le 2P(B_n)+$$

/4.13/ $$+P\Big(\frac{n}{(2+N^+(T))\sum_{k=1}^n H_k(T)}\sup_{-\infty<u\le T}|\sum_{k=1}^n H_k(u)-N^+(u)-2|>\varepsilon,\bar B_n\Big)\le$$

$$\le 2\exp\{-\tfrac{2}{9}\alpha_n^2\}+P\Big(\frac{n}{\mu_n(\mu_n-\alpha_n\sigma_n)}\sup_{-\infty<u\le T}|\sum_{k=1}^n H_k(u)-N^+(u)-2|>\varepsilon\Big).$$

The second term of the last inequality can be estimated by

/4.14/ $$P\Big(\frac{2n}{\mu_n(\mu_n-\alpha_n\sigma_n)}>\frac{\varepsilon}{2}\Big)+P\Big(\frac{n}{\mu_n(\mu_n-\alpha_n\sigma_n)}\sup_{-\infty<u\le T}|\sum_{k=1}^n H_k(u)-N^+(u)|>\frac{\varepsilon}{2}\Big).$$

From condition $\frac{n^{3/2}}{\mu_n(\mu_n-\alpha_n\sigma_n)}<\varepsilon$ follows that the first term of /4.14/ is equal to zero. A very useful Lemma of R.S. Singh [11] gives an upper bound of the second probability of /4.14/. According to this lemma;

/4.15/ $$P(\sup_{-\infty<u\le T}|\sum_{k=1}^n H_k(u)-N^+(u)|\ge a)\le 4a\exp\{-2(\tfrac{a^2}{n}-1)\}$$

holds for each $n\ge 1$, and $a\ge\sqrt{n}$. Using the condition $\frac{4n^{3/2}}{\mu_n(\mu_n-\alpha_n\sigma_n)}<\varepsilon$ and /4.13/, /4.14/, /4.15/ we obtain that

$$P(\sup_{-\infty<u\le T}|R_{n3}(u)|>\varepsilon)\le 2\exp\{-\tfrac{2}{9}\alpha_n^2\}+$$
$$+2e^2\frac{\varepsilon_n(\mu_n-\alpha_n\sigma_n)}{n}\exp\Big\{-\frac{1}{2}\varepsilon^2\frac{\mu_n^2(\mu_n-\alpha_n\sigma_n)^2}{n^3}\Big\}$$

/ii/ To prove /ii/ let us choose $\varepsilon_n=\frac{n^{3/2}\sqrt{2(1+4\log n)}}{\mu_n(\mu_n-\alpha_n\sigma_n)}$. If $n>n_0$ then $\varepsilon_n>\frac{4n^{3/2}}{\mu_n(\mu_n-\alpha_n)}$ and from the first part of the lemma one gets that $\sum_{n=1}^\infty P(\sup_{-\infty<u\le T}|R_{n3}(u)|>\varepsilon_n)<+\infty$.

Thus the statement follows from the Borel-Cantelli lemma.

## 5. Proof of Theorems

**Proof of Theorem 3.1.** Using the wellknown inequality

/5.1/ $$|x-y|\le|\log x-\log y|\qquad\text{if } 0<x\le 1,\ 0<y\le 1,$$

we get that

/5.2/ $$\sup|F_n^*(u)-F(u)|\le\sup|F_n^*(u)-\bar F_n(u)|+\sup|\log\bar F_n(u)-\log F(u)|.$$

Applying the decomposition /4.1/ of $\log\bar F_n(u)$ it follows that

$$\text{/5.3/} \quad \begin{aligned}\sup|\log \bar{F}_n(u) - \log F(u)| &\le \sup|R_{n,1}(u) - \log F(u)| + \\ &+ \sup|R_{n,2}(u)| + \sup|R_{n,3}(u)|\end{aligned}$$

where $R_{n,1}$, $R_{n,2}$ and $R_{n,3}$ are defined by formulae /4.1/- /4.4/ in section 4. Therefore using /5.2/ and /5.3/ and applying assertion /i/ of Lemmas 4.2-4.5 we get that if

$$0 < \max\left\{\frac{n}{(\mu_n - \alpha_n \sigma_n)^2}, \frac{4n^{3/2}}{\mu_n(\mu_n - \alpha_n \sigma_n)}\right\} < \varepsilon$$

then /3.1/ follows. Hence it is enough to show that

$$\text{/5.4/} \quad \frac{n}{(\mu_n - \alpha_n \sigma_n)^2} \le \frac{4n^{3/2}}{\lambda \mu_n(\mu_n - \alpha_n \sigma_n)} .$$

By /iv/ $\alpha_n \sigma_n \le \sigma_n^2$ thus

$$\text{/5.5/} \quad \frac{1}{\mu_n - \alpha_n \sigma_n} \le \frac{1}{\mu_n - \sigma_n^2} = \frac{1}{\sum_{k=1}^{n} H_k^2(T)} .$$

Now /5.4/ follows from inequality

$$\text{/5.6/} \quad \sum_{k=1}^{n} H_k(T) \le \sqrt{n \sum_{k=1}^{n} H_k^2(T)} .$$

and condition /iv/.

Proof of Theorem 3.2. The proof is a direct consequence of inequalities /5.2/, /5.3/ and of parts /ii/ of Lemmas 4.2- 4.5 and of the inequalities that

$$\frac{\sqrt{n \log n}}{\mu_n} \le \frac{n^{3/2}\sqrt{\log n}}{\mu_n(\mu_n - \alpha_n \sigma_n)}, \quad \frac{n}{(\mu_n - \alpha_n \sigma_n)^2} \le \frac{n^{3/2}\sqrt{\log n}}{\mu_n(\mu_n - \alpha_n \sigma_n)}$$

The last inequality follows from inequalities /5.5/ and /5.6/ and the boundedness of $\dfrac{1}{\sum_{k=1}^{n} H_k^2(T)}$ .

## 6. Results for the Bayesian estimator

In this section an improved version of a lemma of Phadia and Van Ryzin [9] is given. It implies that all of the results of section 3 concerning the PL estimator hold for the Bayesian estimator too, in case of nonnegative random variables. Naturally for results involving the interval $[0,T]$ it should be supposed that $\alpha(T) > 0$ .

The following notations and assumptions are needed in the sequel. Let $\alpha$ be an arbitrary non-null positive measure on the Borel $\sigma$-field of $(0,\infty)$ . Assume that

/6.1/ $\alpha(T) = \alpha([T,\infty)) > 0.$

Definition. If $1-F$ is assumed to be a random distribution function with Dirichlet process prior with parameter $\alpha$ , the Bayes estimator of $F(u)$ has been obtained by Susarla and Van Ryzin [12] as

/6.2/ $$F_n^{\alpha}(u) = \frac{N^+(u)+\alpha(u)}{n+\alpha(0)} \prod_{i=1}^{n} \left( \frac{N^+(Z_i)+\alpha(Z_i)+1}{N^+(Z_i)+\alpha(Z_i)} \right)^{[\delta_i=0,\, Z_i \le u]}$$

where $\alpha(u) = \alpha([u,\infty))$, $\alpha(Z_i) = \alpha([Z_i, +\infty))$.

Denote by $Z_{(1)} \le Z_{(2)} \le \dots \le Z_{(n)}$ the ordering of the sample $Z_1, Z_2, \dots Z_n$ and by $\delta_{(i)}$ $(i = 1, 2, \dots n)$ the $\delta$ corresponding to $Z_{(i)}$ in the original sample. Then the product limit estimator may be written as

/6.3/ $$F_n^*(u) = \begin{cases} 1 & \text{if } 0 \le u \le Z_{(i)} \\ \dfrac{N^+(u)}{n} \displaystyle\prod_{j=1}^{i} \left( \frac{N^+(Z_j)+1}{N^+(Z_j)} \right)^{[\delta_{(j)}=0]} & \text{if } Z_i \le u < Z_{(i+1)} \\ & \text{if } Z_n \le u . \end{cases}$$

Lemma 6.1. Suppose that the distribution functions $1-F$, $1-G_1, 1-G_2, \dots 1-G_n$ of the nonnegative random variables $X, Y_1, Y_2, \dots, Y_n$ are continuous on $[0,T]$ moreover $\alpha(T) > 0$ and if $n > n_0$, $\mu_n(T) > 0$ . Let $\alpha_n$ be an arbitrary nonnegative sequence such that $0 \le \alpha_n \le \sigma_n$

/i/ If $n$ and $\varepsilon$ are such that $0 < \dfrac{4n\alpha(0)}{(\mu_n - \alpha_n \sigma_n)^2} < \varepsilon$

$$P(\sup_{0 \le u \le T} |F_n^*(u) - F_n^{\alpha}(u)| > \varepsilon) \le 3 \exp\left(-\frac{2}{9} \alpha_n^2\right).$$

/ii/ If $\sum_{n=1}^{\infty} \exp\left(-\frac{2}{9}\alpha_n^2\right) < +\infty$ then

$$\sup_{0\le u\le T} |F_n^*(u) - F_n^\alpha(u)| = O\left(\frac{n}{(\mu_n - \alpha_n \sigma_n)^2}\right) \quad \text{a.s.}$$

Proof. For $Z_{(i)} \le u < Z_{(i+1)}$ $i = 0, 1, \dots, n-1$ with

/6.4/ $$|F_n^*(u) - F_n^\alpha(u)| = \left| \frac{n-i}{n} \prod_{j=1}^{i} A_j - \frac{n-i+\alpha(u)}{n+\alpha(0)} \prod_{j=1}^{i} B_j \right|$$

where

/6.5/ $$A_j = \left(\frac{n-j+1}{n-j}\right)^{[\delta_{(j)}=0]} \quad \text{and} \quad B_j = \left(\frac{n-j+\alpha(Z_{(j)})+1}{n-j+\alpha(Z_{(j)})}\right)^{[\delta_{(j)}=0]}$$

Observe that

/6.6/ $$|F_n^*(u) - F_n^\alpha(u)| \leqq \left| \frac{n-i}{n} \left( \prod_{j=1}^{i} A_j - \prod_{j=1}^{i} B_j \right) + \prod_{j=1}^{i} B_j \right| \frac{n-i}{n} - \frac{n-i+\alpha(u)}{n+\alpha(0)} \Big| .$$

Using the facts that

/6.7/ $$\frac{n-i}{n} \prod_{j=1}^{i} A_j \le 1 \quad \text{and} \quad \frac{n-i+\alpha(u)}{n+\alpha(0)} \prod_{j=1}^{i} B_j \le 1$$

for $i = 0, 1, \dots n-1$ and $B_j \le A_j$ for $j = 1, 2, \dots i$,

it is easy to see that, as

$$\prod_{j=1}^{n} A_j - \prod_{j=1}^{n} B_j = \sum_{j=1}^{i} \left( \prod_{l=1}^{j-1} B_l \right) (A_j - B_j) \left( \prod_{k=j+1}^{i} A_k \right)$$

/6.8/

$$\left| \frac{n-i}{n} \left( \prod_{j=1}^{n} A_j - \prod_{j=1}^{n} B_j \right) \right| \le \sum_{j=1}^{i} |A_j - B_j| \le \sum_{j=1}^{i} \frac{\alpha(Z_{(j)})}{(n-j)(n-j+\alpha(Z_{(j)}))} \le \frac{i\alpha(0)}{(n-i)^2} .$$

Moreover using /6.7/

/6.9/ $$\prod_{j=1}^{i} B_j \left| \frac{n-i}{n} - \frac{n-i+\alpha(u)}{n+\alpha(0)} \right| \le \frac{\alpha(0)}{n-i} .$$

Consequently for $Z_{(i)} \le u < Z_{(i+1)}$ $(i = 0, 1, \dots n-1)$ from /6.8/ and /6.9/

/6.10/ $$|F_n^*(u) - F_n^\alpha(u)| \le \frac{i\alpha(0)}{(n-i)^2} + \frac{\alpha(0)}{n-i} .$$

Denote by $C_n$ the following sets:

/6.11/ $$C_n = \left\{ \omega : \max_{1\le j\le n} Z_j(\omega) \ge T \right\}$$

According to /6.10/ for $\omega \in C_n$

/6.12/ $$\sup |F_n^*(u) - F_n^\alpha(u)| \le \alpha(0)\left\{\frac{n}{(N^+(T))^2} + \frac{1}{N^+(T)}\right\}$$
$$\le 2\alpha(0)\frac{n}{(N^+(T))^2} \le 4\alpha(0)\frac{n}{(N^+(T)+1)^2},$$

as for $\omega \in C_n$ $N^+(T) \ge 1$.

Taking into account that

/6.13/ $$P(\bar{C}_n) \prod_{k=1}^{n}(1-H_k(T)) \le e^{-\mu_n} \le e^{-\alpha_n^2} \le e^{-\frac{2}{9}\alpha_n^2}$$

both results follow from /6.12/, /6.13/ and Lemma 4.1.

## REFERENCES


[1] Hoeffding, W. Probability inequalities for sums of bounded random variables. J. Amer. Stat. Assoc. 58, 13-30 /1963/.

[2] Burke, M. D. Asymptotic representations of the product limit estimate under random censorship. /to appear/.

[3] Chung, K. L. An estimate concerning the Kolmogorov limit distribution. Trans. Amer. Math. Soc. 67, 36-50 /1949/.

[4] Doob, J. L. Stochastic Processes. J. Wiley and Sons. New York, /1953/.

[5] Dvoretzky, A., Kiefer, J. and Wolfowitz, J. Asymptotic minimax character of the sample distribution function and the classical multinomial estimator. Ann. Math. Stat. 27, 642-669 /1956/.

[6] Földes, A., Rejtő, L. Strong uniform consistency for non-parametric survival estimators from randomly censored data /to appear/.

[7] Földes, A., Rejtő, L. and Winter, B. B. Strong consistency properties of nonparametric estimators for randomly censored data /Part I/. /to appear/.

[8] Kaplan, E. L. and Meier, P. Nonparametric estimation from incomplete observations. J. Amer. Stat. Assoc. 53, 457-481 /1958/.

[9] Phadia, E. G. and Van Ryzin, J. A note on convergence rates for the PL estimator. /to appear/.

[10] Rényi, A. Probability theory. Akadémiai K. Budapest, /1970/.

[11] Singh, R. S. On the Glivenko-Cantelli theorem for weighted empiricals based on independent random variables. Annals of Probability 3, 371-374.

[12] Susarla, V. and Van Ryzin, J. R. Nonparametric Bayesian estimation of survival curves from incomplete observations. J. Amer. Stat. Assoc. 72, 889-902 /1976/.

[13] Susarla, V. and Van Ryzin, J. R. Large sample theory for a Bayesian nonparametric survival curve estimator based on censored samples. Ann. Statist. 6, 755-768 /1978/.

[14] Susarla, V. and Van Ryzin, J. R. Large sample theory for survival curve estimator under variable censoring /to appear/.

[15] Winter, B. B., Földes, A. and Rejtő, L. Glivenko-Cantelli theorems for the PL estimate. Problems of Control and Inf. Theory 7, 213-225 /1978/.

[16] Földes, A. and Rejtő, L. A LIL type result for the product limit estimator on the whole line /to appear/

# A LOCAL CENTRAL LIMIT THEOREM ON SOME GROUPS

PETER GERL (SALZBURG)

## 1. THE PROBLEM

Let G be a locally compact group, p and q probability measures on G. As usual the convolution product $p*q$ is defined by

$$p*q(f) = \int_G \int_G f(xy)\, dp(x)\, dq(y)$$

(f is continuous with compact support on G ) and $p*q$ is again a probability measure on G. In particular we denote the convolution powers of the probability measure p by

$$p^1 = p, \quad p^n = p*p^{n-1}.$$

DEFINITION: We say that there is a local central limit theorem for p if there exist a sequence $(a_n)$ of positive real numbers and a nonzero Radon measure m such that

$$a_n\, p^n \xrightarrow[n \to \infty]{\text{vaguely}} m$$

(i.e. $a_n p^n(f) \to m(f)$ for all real valued continuous functions f on G with compact support).

There is some interest in finding local central limit theorems and to determine how the sequence $(a_n)$ depends on the probability p and on the group G.

In the special case where G is a discrete group, a probability p on G is just a real valued function p on G such that $p(x) \geq 0$ for all $x \varepsilon G$ and $\sum_{x \varepsilon G} p(x) = 1$. Convolution of two probabilities p and q is given by the formula $p*q(x) = \sum_{y \varepsilon G} p(y) q(y^{-1}x)$ $(x \varepsilon G)$. A local central limit theorem for p then asks for a sequence $(a_n)$ and a nonnegative function m on G $(m \neq 0)$ such that $\lim_{n \to \infty} a_n p^n(x)$ ex. $= m(x)$ for all $x \varepsilon G$; i.e. we want to find out the asymptotic behaviour of the convolution powers $p^n$.

If we specialize further and assume that the discrete group G is finitely generated by a set $A \subset G$ (this means that $G = \bigcup_{n=1}^{\infty} A^n$ where $A^n = \{a_1 \ldots a_n \mid a_i \in A\}$) and that p is the uniform distribution on A then we get for $x \in G$

$$p^n(x) = \frac{1}{|A|^n} \text{ (number of representations of } x \text{ as a product of } n \text{ elements from } A)$$

($|A|$ = cardinal number of A). This shows that the convolution powers $p^n$ of a particular probability measure p on G are directly related to the group structure of G.

## 2. SOME RESULTS

For the following results we will always assume:

1) the semigroup generated by the support of p (supp(p)) is dense in G (p adapted)

2) supp(p) is not contained in any coset of a proper closed normal subgroup of G (p aperiodic; for discrete G this implies that to every $x \in G$ there is an integer $n(x)$ such that $p^n(x) > 0$ for all $n \geq n(x)$).

For p a probability measure on a locally compact group G we introduce the notation

$$r_p = \limsup_{n \to \infty} (p^n(V))^{1/n}$$

where V is an open relatively compact neighborhood of e; $r_p$ is independent of V ([12]) and is called the spectral radius of p ($0 \leq r_p \leq 1$).

Under the assumptions 1) and 2) we have the following two general results:

If G is compact then

$$p^n \xrightarrow[n \to \infty]{\text{vaguely}} m = \text{normalized Haar measure on } G,$$

i.e. $a_n = 1$ for all n ([15]).

If G is locally compact, not compact (and with a countable base for the topology) then

$$\frac{p^n}{r_p^n} \xrightarrow[n \to \infty]{\text{vaguely}} 0,$$

i.e. $a_n r_p^n \to \infty$ ([8] for discrete G, [20] in general; weaker results are in [5],[16]).

More detailed results are known in the following cases ($a_n$ refers to the sequence of real numbers of a local central limit theorem of p):

| group G | conditions on p | $a_n$ | reference |
|---|---|---|---|
| $R^k$ or $Z^k$ | moment condition | $n^{k/2} r_p^{-n}$ | [11],[18] |
| motion group of $R^k$ | moment condition | $n^{k/2} r_p^{-n}$ | [1] |
| semisimple connect. Lie group with finite center | "rotation-invariant" | $n^{a/2} r_p^{-n}$ (*) | [3] |
| free group (finitely generat.) | isotropic (p(x) depends only on the "length" of $x \varepsilon G$) | $n^{3/2} r_p^{-n}$ | [17] |
| $\langle a,b \mid \ \rangle$ = free group with 2 gener. | $supp(p) = \{a,b,a^{-1},b^{-1}\}$ | $n^{3/2} r_p^{-n}$ | [9] |
| $\langle a,b \mid a^k = b^k = e \rangle$ | p=uniform distribution on the set $\{a,a^2,\ldots,a^{k-1},b,\ldots,b^{k-1}\}$ | $n^{1/2}$ for k=2; $n^{3/2} r_p^{-n}$ for k≥3 | proof in this paper |

(*) = a is a positive integer determined by G and independent of p

We will prove here the following more precise result which gives also explicitely the limiting measure m, namely the

THEOREM: Let the group $G_k = \langle a,b \mid a^k = b^k = e \rangle$, $H_k = \{a,a^2,\ldots,a^{k-1}, b,b^2,\ldots,b^{k-1}\} \subset G_k$ and $p_k$ be the uniform distribution on $H_k$ (k=2,3,..). Then we have for

1) k=2

$$p_2^{2n}(w) \sim p_2^{2n+1}(w_1) \sim (\pi n)^{-1/2} \quad \text{for } n\to\infty,$$

where w ($w_1$) is a product of an even (odd) number of elements of $H_2$

2) k≥3

$$p_k^n(w) = h_k(w)\, n^{-3/2} r_{p_k}^{\ n} + O(n^{-2} r_{p_k}^{\ n}) \quad \text{for } n\to\infty$$

for all elements $w \varepsilon G_k$; $h_k(w)$ is given explicitely by (20) and

$$r_{p_k} = \lim_{n\to\infty} (p_k^n(e))^{1/n} = \frac{k-2+2(k-1)^{1/2}}{2(k-1)} .$$

## 3.PROOF OF THE THEOREM

The proof will be given in several steps (the method is the same as in [7] and[9]): We first derive equations for several generating functions, then we study these generating functions (singularities, behaviour around these singularities) and finally we apply the method of Darboux to get the desired asymptotic behaviour of $p_k^n$.

a) We fix an integer $k\geq 2$ and write simply $H$ for $H_k$ and $p$ for $p_k$. We draw the graph of the group $G=G_k$, generated by $H$ and indicate by oriented edges the possible transitions given by the probability $p$ (all edges have probability $\frac{1}{2(k-1)}$, since $p$ is the uniform distribution on $H$):

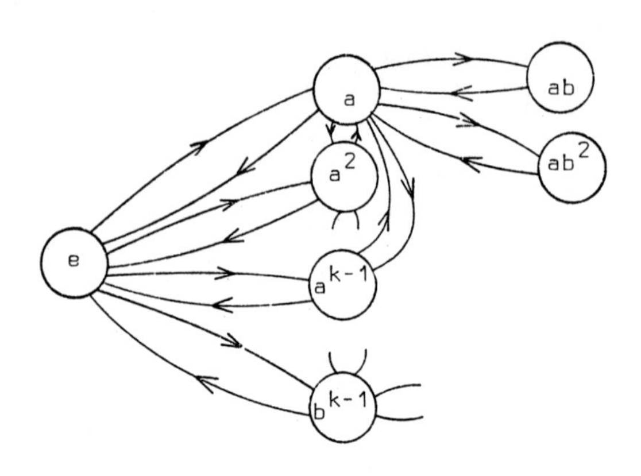


Fig. 1

We will treat at first the case of $p^n(e)$; to do this we can simplify the graph of $G$ by grouping the group elements as follows (mergeable process in the terminology of [14]):

$(0) = \{e\} = H^0$

$(1) = \{a,a^2,\ldots,a^{k-1},b,b^2,\ldots,b^{k-1}\} = H^1$

$(2) = \{ab,ab^2,\ldots,ab^{k-1},a^2b,\ldots,b^{k-1}a^{k-1}\} = H^2 \setminus (H^0 \cup H^1)$

......

This gives the following random walk on the nonnegative integers:

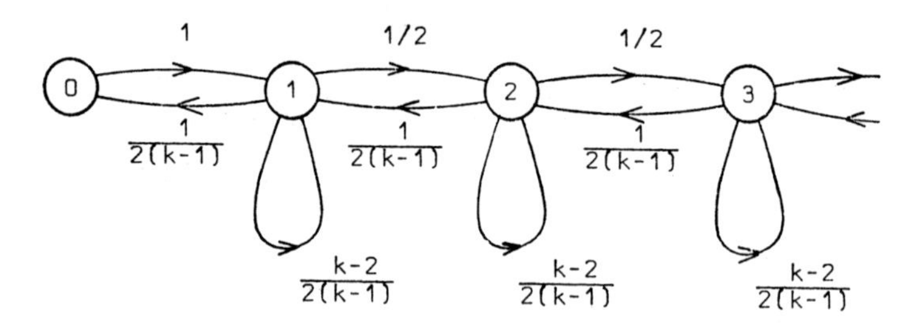


Fig. 2

Now let

$$(1)\quad C = C(z) = \sum_{n\geq 0} p^n(e)z^n = 1 + \frac{1}{2(k-1)}z^2 + \frac{k-2}{4(k-1)^2}z^3 + \cdots$$

$$(2)\quad A = A(z) = \sum_{n\geq 2} a_n z^n = \frac{1}{4(k-1)}z^2 + \frac{k-2}{8(k-1)^2}z^3 + \cdots,$$

where $a_n$ = probability of returning to (1) for the first time after n steps in the random walk of Fig. 2 if we start in (1) and if the first step is the transition (1) → (2).

($p^n(e)$ is also the probability of returning to (0) after n steps in the random walk of Fig. 2 if we start at (0)).

Then we have the following relations by flow graph analysis (see e.g. [14]):

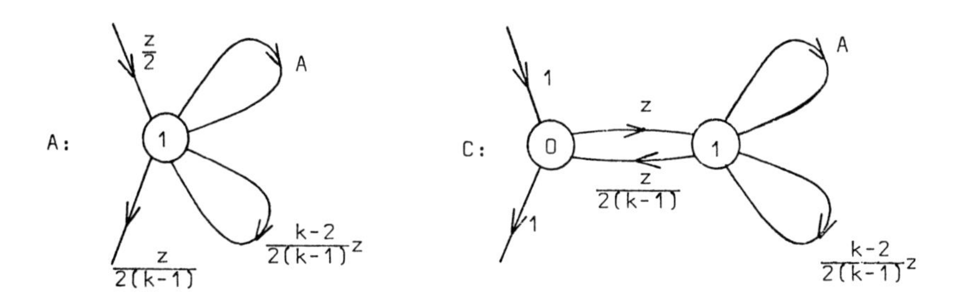


$$(3)\quad A = \frac{z^2}{4(k-1)}\;\frac{1}{1 - A - \frac{k-2}{2(k-1)}z} \quad \text{and}$$

$$(4)\quad C = \frac{1}{1 - \frac{z^2}{2(k-1)}\;\frac{1}{1 - A - \frac{k-2}{2(k-1)}z}} = \frac{1}{1-2A}\,.$$

So by (4) $A = \frac{C-1}{2C}$ and substituting this in (3) gives the following equation for C:

$$(5)\quad f(z,C) = (k-1-(k-2)z-z^2)C^2+(k-2)zC-k+1 = 0.$$

b) We now study the function C; C is the solution of (5) with $C(0)=1$. Since C is an algebraic function the possible singularities of C are either the solutions of (coefficient of $C^2$ vanishes)

$$(6)\quad k-1-(k-2)z-z^2 = 0 \quad \text{or of}$$

$$(7)\quad f(z,C) = 0 \quad \text{and} \quad f_C(z,C) = 0.$$

The solutions of (6) are

$$(8)\quad z = 1 \quad \text{or} \quad z = 1-k$$

and (7) gives (after eliminating C)

$$(8k-k^2-8)z^2+4(k-1)(k-2)z-4(k-1)^2 = 0$$

with the roots

$$(9) \qquad z_1 = \frac{2(k-1)}{k-2+2(k-1)^{1/2}} \quad \text{or} \quad z_2 = \frac{2(k-1)}{k-2-2(k-1)^{1/2}} .$$

Since all the coefficients in (1) are positive a well known theorem of Pringsheim implies that the radius of convergence R ($\geq 1$) of the power series (1) is a singularity of $C(z)$ ($z=R$ is the singularity of $C(z)$ which is closest to the origin), so $R = \limsup (p^n(e))^{-1/n}$ has to be one of the numbers in (8) or (9). If $k \geq 3$ the group $G=G_k$ is not amenable ([6]) and therefore ([2]) $z=1$ cannot be a singularity. Since $1-k<0$ (we assumed $k \gtrsim 2$) R has to be one of the numbers in (9) (for $k \geq 3$). Now $z_1>0$ for all $k \geq 2$ and either $z_2<0$ or $z_2>z_1$. Since the coefficient of $C^2$ in (5) does not vanish for $z=z_1$, the function $C(z)$ is bounded and continuous at $z=z_1$ and the power series (1) converges there (the partial sums are increasing and bounded); therefore a theorem of Abel implies that

$$(10) \qquad C(z_1) = \sum_{n\geq 0} p^n(e)z_1^{\,n} = (\text{by } (5)) = 1 + 2\frac{(k-1)^{1/2}}{(k-2)}.$$

Since $p^n(e)>0$, the power series (1) is absolutely convergent for all z with $|z| \leq z_1$. Now a calculation gives $f_z(z_1,C(z_1)) \neq 0$, $f_C(z_1,C(z_1))=0$ which implies that the branch of $C(z)$ in which we are interested has a vertical tangent at $z=z_1$. Therefore

$$(11) \qquad R = \limsup_{n\to\infty} (p^n(e))^{-1/n} = z_1 = \frac{2(k-1)}{k-2+2(k-1)^{1/2}} = r_p^{\;-1}.$$

Summarizing we found: For $k \geq 3$ the function $C(z)$ defined by (1) has radius of convergence R given by (11); the power series (1) converges absolutely for all z with $|z| \leq R$ and the only singularity on the circle of convergence is $z=R=z_1$.

c) We assume here $k \geq 3$ and derive the asymptotic behaviour of $p^n(e)$ for $n\to\infty$. We saw in b) that $f(z_1,C(z_1))=0$, $f_C(z_1,C(z_1))=0$. A short calculation shows further that $f_z(z_1,C(z_1)) \neq 0$, $f_{CC}(z_1,C(z_1)) \neq 0$, so z can be considered as an analytic function of C in a neighborhood of $(z_1,C(z_1))$ and

$$\frac{dz}{dC}(C(z_1)) = 0, \quad \frac{d^2z}{dC^2}(C(z_1)) \neq 0;$$

therefore the function $C=C(z)$ has a branching point of order 1 at $z=z_1$, i.e. can be developped around $z_1$ in a power series in $(z_1-z)^{1/2}$ (Puiseux-series). Thus we get

$$(12) \qquad C(z) = c_o - c_1(z_1-z)^{1/2} + \ldots$$

Of course $c_o=C(z_1)$ which is given by (10). We next show $c_1 \neq 0$. For

$c_1=0$ would imply $C(z)=c_0+c_2(z_1-z)+\dots$ and $C'(z_1)=-c_2$ would exist. But from (5) we get

$$(13)\qquad \frac{d}{dz}f(z,C(z)) = (-(k-2)-2z)C^2+2(k-1-(k-2)z-z^2)CC'+(k-2)C+(k-2)zC' = 0$$

and using $f_C(z_1,C(z_1))=0$ and (10) we arrive at a contradiction. Therefore $c_1\neq 0$. Since $C(z)$ is strictly increasing (all $p^n(e)\geq 0$), $C'(z)>0$ for $z>0$. From (12) we infer

$$C'(z)(z_1-z)^{1/2} = \frac{c_1}{2} + (z_1-z)^{1/2}(\dots);$$

the left hand side is positive for $0<z<z_1$, therefore $c_1>0$.

To calculate $c_1$ we use (13), $f_C(z_1,(C(z_1))=0$ and $\lim_{z\to z_1} C'(z)(C(z)-C(z_1))=-\frac{1}{2}c_1{}^2$. This gives after some calculation

$$(14)\qquad c_1{}^2 = \frac{c_0((k-2+2z_1)c_0-(k-2))}{(k-2)z_1+z_1{}^2}.$$

Now $C(z_1z)$ has radius of convergence 1 and the only singularity on the unit circle is $z=1$. In a neighborhood of $z=1$ we have from (12) $C(z_1z)=c_0-c_1z_1{}^{1/2}(1-z)^{1/2}+\dots$ Therefore the method of Darboux (see [4] or [19]) gives:

$$p^n(e)z_1{}^n = -c_1z_1{}^{1/2}\binom{1/2}{n}(-1)^n + O(n^{-2}) \quad \text{or}$$

$$(15)\qquad p^n(e) = \frac{c_1z_1{}^{1/2}}{2\pi^{1/2}}\, z_1{}^{-n}\, n^{-3/2} + O(z_1{}^{-n}\, n^{-2}),$$

where $c_1$ is given by (14) and $z_1 = \dfrac{2(k-1)}{k-2+2(k-1)^{1/2}}$ $(k\geq 3)$.

d) The case $k=2$ can be treated directly, namely (5) gives $(1-z^2)C^2-1=0$ or

$$C = (1-z^2)^{-1/2} = \sum_{n\geq 0}\binom{-1/2}{n}(-1)^n z^{2n}.$$

Therefore

$$p^{2n}(e) \sim \pi^{-1/2}\, n^{-1/2}$$

from Stirling's formula (the group $G_2$ is amenable and $z=1$ is a singularity of $C(z)$ as it should be since $p$ is symmetric).

e) If we repeat all the calculations we did with $C(z)$ in a),b),c) with the function $A(z)$ we find: $A(z)$ is the solution of

$$(16)\qquad g(z,A) = 4(k-1)A^2+(2(k-2)z-4(k-1))A+z^2 = 0$$

with $A(0)=0$. For $k\geq 3$ the function $A(z)$ defined by (2) has radius of convergence $R$ given by (11); the power series (2) converges absolutely for all $z$ with $|z|\leq R$ and the only singularity on the circle of convergence is $z=R=z_1$. Around $z=z_1$ we have

$$(17) \qquad A(z) = a_o - a_1(z_1-z)^{1/2} + \ldots$$

where $a_o = A(z_1) = \dfrac{(k-1)^{1/2}}{k-2+2(k-1)^{1/2}}$ and $a_1^2 = \dfrac{(k-2)a_o+z_1}{2(k-1)}$, $a_1>0$.

f) We now treat the case of $p^n(w)$ with $w\varepsilon G=G_k$ (k fixed $\geq 3$). To do this we introduce some more generating functions:

$$C_w(z) = \sum_{n\geq 0} p^n(w)\, z^n \quad (w\varepsilon G), \quad C_e(z) = C(z)$$

$$F_w(z) = \sum_{n\geq 0} f_n(w)\, z^n \quad (w\varepsilon G),$$

where $f_n(w)$ = probability of reaching w for the first time after n steps in the random walk on G(Fig.1) if we start at e.

Then we have the following relations (which can be seen again e.g. by flow graph analysis as in a)): If $w=u_1\ldots u_s \varepsilon G$ with $u_i \varepsilon H$ (the representation of w is reduced), then

$$C_w(z) = F_{u_1}(z)\ldots F_{u_s}(z)C(z), \quad F_u(z) = F_a(z) = \frac{2}{z}A(z) \text{ for } u\varepsilon H.$$

Therefore

$$(18) \qquad C_w(z) = (F_a(z))^s C(z) = (\tfrac{2}{z}A(z))^s C(z),$$

so $C_w(z)$ can be represented as a power series around the origin with radius of convergence R given by (11) and on its circle of convergence is only the singularity $z=R=z_1$ (which is a branching point of order 1). Thus the Puiseux series of $C_w(z)$ around $z_1$ has the form $C_w(z) = c_o(w)-c_1(w)(z_1-z)^{1/2}+\ldots$ This together with (18),(17) and (12) implies

$$c_o(w)-c_1(w)(z_1-z)^{1/2}+\ldots = \frac{2^s}{z_1^s}\left(1+\frac{z_1-z}{z_1}+\ldots\right)^s (a_o-a_1(z_1-z)^{1/2}+\ldots)^s (c_o-c_1(z_1-z)^{1/2}+\ldots)$$

and coparison of the coefficients yields

$$(19) \qquad c_o(w) = \frac{2^s}{z_1^s} a_o^s c_o, \quad c_1(w) = \frac{2^s}{z_1^s}(a_o^s c_1 + s c_o a_o^{s-1} a_1).$$

Another application of the method of Darboux then gives the result

$$p^n(w) = \frac{c_1(w) z_1^{1/2}}{2\pi^{1/2}} z_1^{-n} n^{-3/2} + O(z_1^{-n} n^{-2}) \qquad (k\geq 3);$$

thus with the notation of the theorem

$$(20) \qquad h_k(w) = \frac{z_1^{1/2}}{2\pi^{1/2}} c_1(w),$$

where $c_1(w)$ is given by (19).

The case k=2 can be treated in a similar way.

## 4. SOME CONJECTURES

From all the results listed in section 2 the following conjecture seems reasonable:

CONJECTURE 1: Let G be a group and p an adapted, aperiodic probability measure on G. Then there exist a positive integer a and a non-zero Radon measure m such that

(21) $$n^{a/2} r_p^{-n} p^n \xrightarrow[n\to\infty]{\text{vaguely}} m.$$

$r_p$ is the spectral radius of p and the integer a depends only on the group G (and not on p), but how?

If we consider the left hand side of (21) as a function of p we get the following weaker

CONJECTURE 2: Let G be a group and p,q adapted, aperiodic probability measures on G. Is it true that

$$\lim_{n\to\infty} \frac{r_p^{-n} p^n(f)}{r_q^{-n}\, q^n(f)} \quad \text{exists and is} \neq 0$$

for all continuous, real valued functions f on G with compact support?

If we fix the probability measure p in (21) then conjecture 1 implies the following weaker

CONJECTURE 3: Let G be a group and p an adapted, aperiodic probability measure on G. Is it true that

$$\lim_{n\to\infty} \frac{p^n(f)}{p^n(g)} \quad \text{exists}$$

for all continuous, real valued functions f,g on G with compact support? What is the value of this limit? (A statement of this kind is called a strong ratio limit theorem, see [13]).

Finally we formulate

CONJECTURE 4: Let $F_2 = \langle a,b | \rangle$ be the free group on two generators and let p be an adapted, aperiodic probability measure on $F_2$. Does one have a local central limit theorem for p with

$$a_n = r_p^{-n}\, n^{3/2} \text{ ?}$$

(see [9],[10],[17] for special cases).

## REFERENCES


1. Baldi P.,Bougerol P.,Crepel P.: Théorème central limite local sur les extensions compactes de $R^d$. Ann.Inst.H.Poincaré. Sect.B (N.S.) 14(1978), 99-112
2. Berg C.,Christensen J.P.R.: Sur la norme des opérateurs de convolution. Inventiones math. 23(1974), 173-178
3. Bougerol P.: Comportement asymptotique des puissances de convolution d'une probabilité sur un espace symétrique. (preprint)

4. Comtet L.: Advanced combinatorics.D.Reidel Publishing Company, Dordrecht-Boston (1974)
5. Derriennic Y.: Lois "zéro ou deux" pour les processus de Markov. Applications aux marches aléatoires. Ann.Inst.H.Poincaré,Sect.B (N.S.) 12(1976), 111-129
6. Dixmier J.: Les moyennes invariantes dans les semi-groupes et leurs applications. Acta Sci.Math.Szeged 12(1950), 213-227
7. Gerl P.: Eine asymptotische Auswertung von Faltungspotenzen. Sitzungsber. Österr.Akad.Wiss.,Math.-naturw.Klasse, Abt. II, 186(1978), 385-396
8. Gerl P.: Ein Konvergenzsatz für Faltungspotenzen.In Probability measures on groups.Lecture Notes in Math. 706 (1979), 120-125
9. Gerl P.: Ein Gleichverteilungssatz auf $F_2$. In Probability measures on groups. Lecture Notes in Math. 706 (1979), 126-130
10. Gerl P.: Asymptotic behaviour of convolution powers on $F_2$. $\pi\alpha\mu$ (Arbeitsber.Math.Inst.Univ.Salzburg) 2(1979), 73-79 (preprint)
11. Gnedenko B.V.,Kolmogorov A.N.: Limit distributions for sums of independent random variables. Addison Wesley, Reading Mass.(1954)
12. Guivarc'h Y.: Loi des grands nombres et rayon spectral d'une marche aléatoire sur un groupe de Lie (preprint)
13. Guivarc'h Y.: Théorêmes quotients pour les marches aléatoires (preprint)
14. Howard R.A.: Dynamic probabilistic systems.Vol.I.John Wiley & Sons,Inc.,New York-London-Sydney-Toronto (1971)
15. Kawada Y., Ito K.: On the probability distribution on a compact group. Proc.Phys.Math.Soc.Japan 22(1940), 977-999
16. Mukherjea A.,Tserpes N.A.: Measures on topological semigroups. Lecture Notes in Math. 547 (1976)
17. Sawyer S.: Isotropic random walks in a tree. Z.Wahrscheinlichkeitstheorie und Verw. Gebiete 42(1978), 279-292
18. Stone Ch.: A local limit theorem for multidimensional distribution functions. Ann.Math.Stat. 36(1965), 546-551
19. Szegö G.: Orthogonal polynomials. American Math. Society, Providence, Rhode Island 1939
20. Woess W.: (Oral communication)

Peter Gerl
Math. Inst. der Universität
Petersbrunnstraße 19
A-5020 SALZBURG
Austria

# ON THE STATISTICS OF GIBBSIAN PROCESSES

Erhard Glötzl
Berthold Rauchenschwandtner

Universität Linz

## 1) Definition and examples of Gibbsian processes

Point processes P are probability measures on the set N of all integer valued Radon measures $\mu$ on a certain phase space T which for most of the applications is $\mathbb{Z}^d$ (lattice models) or $\mathbb{R}^d$ (continuous models). The name "point process" comes from the fact, that an integer valued Radon measure can be identified with a corresponding configuration of points in the phase space.

The concept of Gibbsian processes as a special class of point processes arose in statistical mechanics to describe systems of interacting particles. The interaction is usually defined in terms of specifications, conditional energies, local energy, or potentials. A survey of the various definitions of Gibbsian processes one can find in Glötzl [4].
It turns out that the conditions under which a general point process can be described as Gibbsian point process are very weak.

The most famous example of a Gibbsian process is the Ising model on $\mathbb{Z}^2$. Let us interprete the Ising model as a lattice gas and write $\mu(x) = 1$ or $0$ according wether there is a point on $x \in \mathbb{Z}^2$ or not and denote the four nearest neighbour lattice sites of x by $x_i$ $(i = 1,2,3,4)$, then the conditional probability

$$P(\mu(x) = s_o \mid \mu(x_i) = s_i) = \frac{1}{Z} \exp(s_o(\beta . \Sigma s_i + \alpha)) \qquad (s_o, s_i \in \{0,1\})$$

where $\alpha \in \mathbb{R}$ is called the chemical potential, $\beta \in \mathbb{R}$ is called the inverse temperature and the normalizing factor Z is called the partition function.

In a more general case the point process P is called Gibbs process with respect to the local energy E, if the conditional probability $P(\delta_x/\mu)$, which is the probability that there is a point in $x \in T$ under the condition that the configuration $\mu$ is known, is given by

$$P(\delta_x/\mu) = \frac{1}{Z} \exp(-E(x,\mu))Q(\delta_x)$$

where $E(x,\mu)$ is called local energy and the "weight process" Q is a fixed point process with independent increments, which in continuous models usually is the Poisson process and in lattice models the Bernoulli process. Z is again the normalizing factor. Note that the above intuitive definition is not exact for continuous models, because $\{\delta_x\}$ is i.g. a zero event. An exact definition in the general case using "local specifications" can be found in Preston [6].

In most of the applications (in $T = \mathbb{Z}^d$ or $\mathbb{R}^d$) the local energy is generated by a pair potential $\phi : [0,\infty[ \rightarrow ]-\infty,\infty]$ and a chemical potential $\alpha \in \mathbb{R}$. In this case

(1) $$E(x,\mu) := \int \phi(\|x-y\|)\mu(dy) + \alpha$$

$\phi$ is said to have finite interaction range $R \in \mathbb{R}$ if $\phi(r) = 0$ for all $r \geq R$ and hard core range $R^*$ if $\phi(r) = \infty$ for all $0 \leq r \leq R^*$.

Examples:

(a) If $\phi \equiv 0$ and $\alpha = 0$ then $P \equiv Q$

(b) In the lattice case if

$$\phi(r) = \begin{cases} \infty & \text{for } r = 0 \\ -\beta & \text{for } r = 1 \\ 0 & \text{otherwise} \end{cases}$$

we get the Ising model.

(c) In the continuous case if

$$\phi(r) = \begin{cases} \infty & \text{for } 0 \le r \le R^* \\ 0 & \text{otherwise} \end{cases}$$

we get the hard core Poisson model.

## 2) Statistics of point processes

In the statistics of point processes one proceeds as follows:

a) assume that the underlying point process is stationary
b) observe one point configuration $\psi$
c) get an estimation for the
   α) first and second moment and related questions
   β) distribution of the distance of the nearest neighbours
   γ) parameters of a certain underlying model (e.g. linear model or Gibbsprocess with pair potential)

One method for the last question c)γ) is due to Besag [1], which is based on coding patterns and maximum likelihood estimations. We propose another method which is especially suitable for Gibbsian processes with pair potentials.

## 3) Theory

Let P be a stationary Gibbsian point process on $\mathbb{R}^d$ or $\mathbb{Z}^d$ with respect to a local energy E and let $\lambda$ be the intensity measure of the "weight process" Q (usually some multiple of the Lebesgue measure on $\mathbb{R}^d$ or the counting measure on $\mathbb{Z}^d$).

<u>Proposition 1</u>: (Georgii [2], Glötzl [4], Rauchenschwandtner [7])
Let P be a Gibbsian point process w.r.t. the local energy E then

$$(2) \qquad E(x,\mu) = -\log \frac{dC_P^!}{d(\lambda \times P)}(x,\mu) \quad \text{for } (\lambda \times P)\text{-a.a. } (x,y)$$

where $C_P^!$ is the reduced Campbell measure defined by

$$C_P^!(B\times F) = \int_N \int_T i_{B\times F}(x, \mu-\delta_x)(dx)P(d\mu)$$

(B a bounded Borel subset of T, F a measurable set of configurations).

Now for simplicity let us assume $T = \mathbb{R}^2$.

Let R denote the range of interaction and for all natural numbers k,n,m all integers i,j and all configurations $\mu,\psi$ let

$$K_R := \{y \mid y \in \mathbb{R}^2, \ 0 \le \|y\| \le R\}$$

$$Q_{i,j}^m := [\tfrac{i}{m}, \tfrac{i+1}{m}[ \times [\tfrac{j}{m}, \tfrac{j+1}{m}[$$

$$A^m(\mu) := \{\varphi \mid \varphi(Q_{i,j}^m) = \mu(Q_{i,j}^m) \text{ for all } i,j \text{ with } Q_{i,j}^m \cap K_R = \emptyset\}$$

$$W_n^m := \{(\tfrac{i}{m}, \tfrac{j}{m}) \mid |\tfrac{i}{m}| \le n, \ |\tfrac{j}{m}| \le n, \ i,j \text{ integers}\} \subset \mathbb{R}^2$$

$$\Theta_t\psi := \text{shifting of } \psi \text{ with } t \in \mathbb{R}^2$$

$$F_n^m(\mu,\psi) = \sum_{t\in W_n^m} i_{A^m(\mu)}(\Theta_t\psi)$$

Roughly speaking $F_n^m(\mu,\psi)$ denotes the following:

Shift according to all $t \in W_n^m$ and observe the frequency, how often in the m-th discretisation $\Theta_t\psi$ corresponds with $\mu$ within the range of interaction.

From the following theorem one derives in an obvious way a consistent estimation for the pair potential $\phi$.

<u>Theorem</u>: Let P be a stationary Gibbsian process with pair potential $\phi$ on the phase space $\mathbb{R}^2$, hard core $R^*$ and finite range of interaction R then for P-almost all $\psi$ and P-almost all $\mu$ for all $x \in \mathbb{R}^2$

with $\mu(x) \geq 1$ and all $y \in \mathbb{R}^2$ with $\mu(y) \geq 1$ and $r := \|x-y\| \in ]R^*,R]$

$$\lim_{m\to\infty}\lim_{n\to\infty} \frac{F_n^m(\Theta_{-x}\mu,\psi)F_n^m(\Theta_{-x}(\mu-\delta_x-\delta_y),\psi)}{F_n^m(\Theta_{-x}(\mu-\delta_x),\psi)F_n^m(\Theta_{-x}(\mu-\delta_y),\psi)} = e^{-\phi(r)}$$

Further if $r = R^* = 0$ then for P-almost all $\psi$ and P-almost all $\mu$ for all x with $\mu(x) \geq 2$

$$\lim_{m\to\infty}\lim_{n\to\infty} \frac{F_n^m(\Theta_{-x}\mu,\psi).F_n^m(\Theta_{-x}(\mu-2\delta_x),\psi)}{(F_n^m(\Theta_{-x}(\mu-\delta_x),\psi))^2} \cdot \frac{\mu(x)}{\mu(x)-1} = e^{-\phi(0)}$$

For the proof we need three lemmata:

Lemma 1: Let P be a Gibbsian point process with respect to some local energy E then for all A

$(\lambda \times P)(A) = 0 \implies$

$\implies P(\{\mu \mid \text{there is a x with } \mu(x) \geq 1 \text{ and } (x,\mu-\delta_x) \in A\}) = 0$

Proof: Assume $P(\{\mu|\text{there is a x with } \mu(x) \geq 1 \text{ and } (x,\mu-\delta_x) \in A\}) > 0$ then $C_P^!(A) = \int_N \int_T i_A(x,\mu-\delta_x)\mu(dx)P(d\mu) > 0$
and because of Proposition 1

$C_P^!(A) = \int_A e^{-E(x,\mu)}(\lambda \times P)(dx,d\mu) > 0$

which implies $(\lambda \times P)(A) > 0$.

Lemma 2: Let P be as in Lemma 1 then P has property $\Sigma'$, i.e.

$P(F) = 1 \implies P(\{\mu|(\mu-\delta_x) \in F \text{ for all x with } \mu(x) \geq 1\}) = 1$

Proof: Rauchenschwandtner [7].

Lemma 3: Let P be as in Lemma 1 then

$P(\{\mu|\text{there is a x with } \mu(x) \geq 1 \text{ and } E(x,\mu-\delta_x) = \infty\}) = 0$

<u>Proof</u>: Denote $\varphi_B$ the restriction of $\varphi$ to B and $\varphi_{\bar{B}}$ the restriction of $\varphi$ to the complement of B then (cf. [5],[7],[9]) for all bounded Borel sets and $(Q \times P)$ a.a. $(\varphi,\mu)$ with $\varphi_B = \sum_{i=1}^{n} \delta_{x_i}$ for all permutations $(\pi_1,\dots,\pi_n)$

$$E(x_1,\mu_{\bar{B}}) + E(x_2,\mu_{\bar{B}}+\delta_{x_1}) + \dots E(x_n,\mu_{\bar{B}} + \sum_{i=1}^{n-1} \delta_{x_i}) =$$

$$E(x_{\pi_1},\mu_{\bar{B}}) + E(x_{\pi_2},\mu_{\bar{B}} + \delta_{x_{\pi_1}}) + \dots E(x_{\pi_n},\mu_{\bar{B}} + \sum_{i=1}^{n-1} \delta_{x_{\pi_i}})$$

for which we write shortly $\tilde{E}(\varphi_B,\mu_{\bar{B}})$.

Then with the DLR equation for all B

$$P(\{\mu | \text{there is a } x \in B \text{ with } \mu(x) \geq 1 \text{ and } E(x,\mu-\delta_x) = \infty\}) =$$

$$= \int_N \int_N i_{\{\mu | \text{there is a } x \in B \text{ with } \mu(x) \geq 1 \text{ and } E(x,\mu-\delta_x)=\infty\}}(\mu) \,.$$

$$.Z^{-1}(\mu,B)e^{-E(\varphi_B,\mu_{\bar{B}})} Q(d\varphi)P(d\mu) = 0.$$

from which the lemma follows immediately.

<u>Proof of the Theorem</u>:

Let us abbreviate for all $k,m,n \in \mathbb{N}$ and all configurations $\mu$

$K_{R,k}$ := union of all $Q^k_{i,j}$ with $Q^k_{i,j} \cap K_R = \emptyset$

$A^m_k(\mu) := \{\varphi | \varphi(Q^m_{i,j}) = \mu(Q^m_{i,j})$ for all i,j with $Q^m_{i,j} \cap K_{R,k} = \emptyset\}$

A) Assume P ergodic:

1) For P-a.a. $\mu$ and P-a.a. $\psi$

$$E(0,\mu) = \frac{dC^!_P}{d(\lambda \times P)}(0,\mu)$$

↑ prop.1 + stationarity

$$:= \lim_{m\to\infty} \frac{C^!_P(Q^m_{0,0},A^m_k(\mu))}{\lambda(Q^m_{0,0})P(A^m_k(\mu))} \qquad \text{for all } k \in \mathbb{N}$$

↗ a)

$$\underset{b)}{\nearrow} = \lim_{m\to\infty} \frac{C_P^!(Q_{0,0}^m, A^m(\mu))}{\lambda(Q_{0,0}^m)\cdot P(A^m(\mu))}$$

$$\underset{\text{Def. of } C_P^!}{\nearrow} = \lim_{m\to\infty} \frac{\int_N \int_{Q_{0,0}^m} i_{A^m(\mu)}(\varphi-\delta_x)\,\varphi(dx)\,P(d\varphi)}{\lambda(Q_{0,0}^m)\cdot P(A^m(\mu))}$$

$$= \lim_{m\to\infty} \frac{\int_N \int_{Q_{0,0}^m} i_{A^m(\mu+\delta_0)}(\varphi)\,\varphi(dx)\,P(d\varphi)}{\lambda(Q_{0,0}^m)\cdot P(A^m(\mu))}$$

$$= \lim_{m\to\infty} \frac{P(A^m(\mu+\delta_0))}{P(A^m(\mu))} \cdot \frac{\mu(Q_{0,0}^m)+1}{\lambda(Q_{0,0}^m)}$$

$$\underset{c),d)}{\nearrow} = \lim_{m\to\infty}\lim_{n\to\infty} \frac{F_n^m(\mu+\delta_0,\psi)}{F_n^m(\mu,\psi)} \cdot \frac{\mu(Q_{0,0}^m)+1}{\lambda(Q_{0,0}^m)}$$

a) Because of the wellknown constructive approach to the Radon nikodym derivative (see e.g. Gihman, Skorohod [3], p.64) for all $k \in \mathbb{N}$ since $\frac{dC_P^!}{d\lambda \times P}(0,\mu)$ does not depend how $\mu$ is outside $K_{R,k}$.

b) Since $A_k^m(\mu) = A^m(\mu)$ for $k \geq m$ we can use a diagonal argument.

c) Because of the multidimensional ergodic theorem (cf. e.g. Nguyen, Zessin [10], Theorem 3.7) and the ergodicity of P for all $\varphi$ and P-a.a. $\psi$

$$\lim_{n\to\infty} \frac{F_n^m(\varphi,\psi)}{|W_n|} = P(A^m(\varphi))$$

($|W_n|$ denotes the counting measure of $W_n$)

d) For all $m \in \mathbb{N}$ and P-a.a. $\mu$ $P(A^m(\mu)) > 0$.

2) Because of stationarity and 1) for P-a.a. $\psi$ for all x and P-a.a. $\mu$

$$E(x,\mu) = E(0,\Theta_{-x}\mu)$$

$$= \lim_{m\to\infty}\lim_{n\to\infty} \frac{F_n^m(\Theta_{-x}\mu+\delta_0,\psi)}{F_n^m(\Theta_{-x}\mu,\psi)} \cdot \frac{\Theta_{-x}\mu(Q_{0,0}^m)+1}{\lambda(Q_{0,0}^m)}$$

3) Because of 2) and Lemma 1 for P-a.a. $\psi$ and P-a.a. $\mu$ for all x with $\mu(x) \geq 1$ since $\Theta_{-x}(\mu-\delta_x)+\delta_0 = \Theta_{-x}\mu$ and $\Theta_{-x}(\mu-\delta_x)(Q_{0,0}^m)+1 = \Theta_{-x}\mu(Q_{0,0}^m)$

$$E(x,\mu-\delta_x) = \lim_{m\to\infty}\lim_{n\to\infty} \frac{F_n^m(\Theta_{-x}\mu,\psi)}{F_n^m(\Theta_{-x}(\mu-\delta_x),\psi)} \cdot \frac{\Theta_{-x}\mu(Q_{0,0}^m)}{\lambda(Q_{0,0}^m)}$$

4) Because of 3) and Lemma 2 for P-a.a. $\psi$ and P-a.a. $\mu$ for all x with $\mu(x) \geq 1$ and all y with $\mu(y) \geq 1$

$$E(x,\mu-\delta_x-\delta_y) = \lim_{m\to\infty}\lim_{n\to\infty} \frac{F_n^m(\Theta_{-x}(\mu-\delta_y),\psi)}{F_n^m(\Theta_{-x}(\mu-\delta_x-\delta_y),\psi)} \cdot \frac{\Theta_{-x}(\mu-\delta_y)(Q_{0,0}^m)}{\lambda(Q_{0,0}^m)}$$

5) Because of Lemma 2 and Lemma 3 for P-a.a. $\mu$ for all x with $\mu(x) \geq 1$ and all y with $\mu(y) \geq 1$ $E(x,\mu-\delta_x-\delta_y) < \infty$.

6) From the definition of the pair potential follows: $\phi(r) = E(x,\mu-\delta_x) - E(x,\mu-\delta_x-\delta_y)$ for all $\mu$ for all x,y with $\mu(x) \geq 1$, $\mu(y) \geq 1$ and $E(x,\mu-\delta_x-\delta_y) < \infty$.

7) The theorem follows for an ergodic P immediately from 3), 4), 5), 6).

B) Let P be not ergodic:

Let E be a fixed local energy and $\{P_i \mid i \in I\}$ the set of all ergodic Gibbsian point process with respect to the local energy E. It is well known (cf. e.g. Preston [6]) that for every Gibbsian point process P with respect to the local energy E there exists a unique measure $\Lambda$ such that

$$P = \int_I P_i \, \Lambda(di)$$

Because of A) for all $i \in I$ for $P_i$-a.a. $\psi$ and $\mu$ the limit converges to $e^{-\phi(r)}$. Therefore and because of the ergodic desintegration of P the limit converges to $e^{-\phi(r)}$ for P-a.a. $\psi, \mu$.

Remark:

1) In the case of a lattice phase space the $\lim_{m\to\infty}$ becomes trivial.

2) A consistent estimation for the chemical potential $\psi$ is easily derived from

$$\alpha = E(0, \sum_i \delta_{x_i}) - \sum_i \phi(\|x_i\|)$$

4) Example

To estimate the inverse temperatur $\beta$ in the two dimensional Ising model as described in chapter 1 one proceeds as follows.
For every configuration $\mu$ let $\mu(i,j) = 1$ or 0 wether there is a point on the lattice site i,j or not and let
$N(\mu) = \mu(0,1) + \mu(1,0) + \mu(-1,0) + \mu(0,-1)$ be the number of neighbouring points of (0,0) in $\mu$.
Let $\nu_1$ be such that $\nu_1(0,0) = 0$ and $N(\nu_1) = 1$

Translate the observed configuration $\psi$ in k different ways, let
$F_1(\nu_1)$ denote the frequency that $\Theta_t\psi(0,0) = 1$ and $N(\Theta_t\psi) = 1$
$F_2(\nu_1)$ denote the frequency that $\Theta_t\psi(0,0) = 0$ and $N(\Theta_t\psi) = 0$
$F_3(\nu_1)$ denote the frequency that $\Theta_t\psi(0,0) = 0$ and $N(\Theta_t\psi) = 1$
$F_4(\nu_1)$ denote the frequency that $\Theta_t\psi(0,0) = 1$ and $N(\Theta_t\psi) = 0$

then

$$\beta_1 = \log \frac{F_1(\nu_1)\cdot F_2(\nu_1)}{F_3(\nu_1)\cdot F_4(\nu_1)}$$

is an estimation for $\beta$ (consistent for $k \to \infty$).

In the same way one gets estimations $\beta_2, \beta_3, \beta_4$ for $\nu_i$ such that $\nu_i(0,0) = 0$ and $N(\nu_i) = i$ $(i=2,3,4)$.

These four estimations can be combined to one estimation $\beta$ in a suitable way, we do not discuss further. It turns out that the standard deviation of $\beta$ is not larger than $\frac{1.5}{\sqrt{k}}$.

Remark: Another method for the estimation of $\beta$ in the Ising model one can find in Strauss [8].

## Literature


[1] BESAG J., Spatial interaction and the statistical analysis of lattice systems, J. Royal Stat.Soc.B, 26 (1974), 192-236.

[2] GEORGII, H.O., Canonical and grand canonical Gibbs states for continuous systems, Comm.Math.Phys. 48, 31-51 (1976).

[3] GIHMAN I.I., SKOROHOD A.V., The Theory of Stochastic Processes I, Springer 1974.

[4] GLÖTZL E., Gibbsian description of point processes, Porceedings of the Colloquium on Point Processes and Quening Theory of the Bolyai Janos Mathematical Society, Sept. 1978, Keszthely, North Holland (to appear).

[5] GLÖTZL E., Lokale Energien und Potentiale für Punktprozesse, Mathematische Nachrichten (to appear).

[6] PRESTON C., Random Fields, Lecture Note, Springer (1976).

[7] RAUCHENSCHWANDTNER B., Gibbsprozesse und Papangeloukerne, Dissertation. Linz 1978.

[8] STRAUSS D.J., Analysing binary lattice data with the nearest neighbour property, J. Appl.Prob. 12, 702-712 (1975).

[9] MATTHES S.K., WARMUTH W., MECKE J., Bemerkungen zu einer Arbeit von Nguyen Xuan Xanh und Hans Zessin, Math. Nachr. 88 (1979), 117-127.

[10] NGUYEN X.X., ZESSIN H., Ergodic Theorems for spatial processes, Zeitschrift f. Wahrscheinlichkeitstheorie verw. Gebiete 48 (1979), 133-158.

EFFICIENCY OF ESTIMATES IN NONREGULAR CASES

Wilfried Grossmann

University of Vienna

## 1. Introduction

Let $\{P_\theta\}$ be a family of probability measures on a measurable space $(X,\mathfrak{A})$ indexed by the parameter $\theta$ where $\theta \in \Theta$ and $\Theta$ is an open set in $\mathbb{R}$. We assume that $P_\theta$ is absolutely continuous with respect to some $\sigma$ finite measure $\nu$ on $(X,\mathfrak{A})$ with density function $f(x,\theta)$. $P_\theta^n$ is the independent product of n identical components of $P_\theta$ with density $f_n(\underline{x},\theta)$, $\underline{x} = (x_1,\dots,x_n)$. One possibility to define efficiency of an estimate $T_n$ for $\theta$ is by covering probabilities. One has to select first a class of appropriate estimates and gives then upper bounds for the probability that $T_n$ lies within an interval $(\theta-t_1\delta_n,\theta+t_2\cdot\delta_n)$ where $t_i > 0$ $i = 1,2$ and $\delta_n$ is a norming sequence with $\delta_n \to 0$ which depends on the family $\{P_\theta\}$. This concept was used by Wolfowitz (1974) who defined a class of estimates in which the maximum probability estimate is optimal. Another possibility is to use the method of Bahadur and derive bounds for the class of median unbiased estimates (Pfanzagl (1970)) or more general strongly asymptotic median unbiased estimates (Michel (1978)). This method was only defined for families which are asymptotically normal. The aim of the paper is to show that this method can also be used for nonregular cases where other than normal limit structures of the loglikelihoodratios occur. In section 2 we give a general result which gives upper bounds for the covering probabilities. In section 3 we give applications to asymptotic normal distributed families and in section 4 we treat the case where the densities have discontinuities.

## 2. The main result

Let $d^2(\theta_1,\theta_2)$ be the Hellinger distance between $P_{\theta_1}$ and $P_{\theta_2}$. We assume that the following condition A is fulfilled.

(2.1) Condition A:

(2.1.1) For $\theta_1 \neq \theta_2$ we have $P_{\theta_1} \neq P_{\theta_2}$

(2.1.2) There exists a sequence $\{\delta_n\}$ with $\lim\limits_{n\to\infty} \delta_n = 0$ such that

$$\lim n.d^2(\theta,\theta + t\delta_n) = a(t,\theta) \quad \text{for all } t \in \mathbb{R} \text{ and } \theta \in \Theta$$

Since (2.1) together with a weak regularity condition implies the existence of sequences of estimates $\{T_n\}$ so that $\delta_n^{-1}(T_n-\theta)$ is bounded in probability (Grossmann (1979)), we define strongly asymptotically median unbiased estimates in the following way:

(2.2) Definition:

Let $\{P_\theta\}$ be a family which fulfills (2.1) with some $\{\delta_n\}$ and $a(t,\theta)$. Then a sequence of estimates $\{T_n\}$ is strongly asymptotically median unbiased if for all $\theta \in \Theta$ and all $t_1,t_2 > 0$

$$\underline{\lim}\; P^n_{\theta+t_2\delta_n}\{T_n(\underline{x}) \geqslant \theta + t_2\delta_n\} \geqslant 1/2$$

$$\underline{\lim}\; P^n_{\theta-t_1\delta_n}\{T_n(\underline{x}) \leqslant \theta - t_1\delta_n\} \geqslant 1/2$$

We denote by $\mathcal{M}$ the class of all strongly asymptotically median unbiased estimates.

(2.3) Remark:

(2.2) generalizes the definition of Michel (1978) to more general situations where the loglikelihoodratio is not necessarily asymptotically normally distributed. The class $\mathcal{M}$ contains all estimates which have a distribution function which is continuously convergent to some limit distribution. Continuous convergence is in close connection to the class $\mathcal{W}_r$ of estimates which was defined by Wolfowitz as the class of all estimates $\{T_n\}$ so that

$$\lim_{n\to\infty}[P^n_{\theta+t\delta_n}\{\delta_n^{-1}(T_n - \theta - t\delta_n) \in (-r,r)\} -$$
$$-P^n\{\delta_n^{-1}(T_n-\theta) \in (-r,r)\}] = 0$$

In $\mathcal{W}_r$ the maximum probability estimate $M_n(\underline{x})$ which is defined as that value d for which

$$\int_{d-r\delta_n}^{d+r\delta_n} f_n(\underline{x},\theta)d\theta$$

attains its maximum. $M_n$ is optimal in that sense that the probability $P^n_\theta\{\theta - r\delta_n < T_n < \theta + r\delta_n\}$ is maximized in $\mathcal{W}_r$ by taking for $T_n$ the maximum probability estimate $M_n$. However as we will see in section 3 and 4 efficiency in $\mathcal{M}$ and $\mathcal{W}_r$ coincide only in the regular case of local asymptotic normality.

In order to formulate our theorem we introduce the following notation. Let $l_n(t,\theta) = f_n(\underline{x},\theta + t\delta_n)/f_n(\underline{x},\theta)$ and let $F^o_n(\cdot;t,\theta)$ and $F^1(\cdot;t,\theta)$ be the distribution functions belonging to $\mathcal{L}(l_n(t,\theta)|P^n_\theta)$ and $\mathcal{L}(l_n(t,\theta)|P^n_{\theta+t\delta_n})$ respectively.

(2.4) Assumptions:

(2.4.1) $\mathcal{L}(l_n(t,\theta)|P_\theta)$ and $\mathcal{L}(l_n(t,\theta)|P^n_{\theta+t\delta_n})$ converge weakly to some limit distributions with distribution functions $F^o(\cdot;t,\theta)$ and $F^1(\cdot;t,\theta)$ for all t and $\theta$.

(2.4.2) Let

$$K(t) = \inf\{K|F^1(K;t,\theta) \geq 1/2\}$$

Then for all $|\varepsilon| < \varepsilon_0$ $K(t) + \varepsilon$ is continuity point of $F^i(\cdot;t,\theta)$ $i = 0,1$ and $F^1(K(t) + \varepsilon;t,\theta) \neq 1/2$.

(2.5) Theorem:

Let (2.1) and (2.3) be fulfilled and let $t_1 < 0$ and $t_2 > 0$. Then for every sequence of strongly asymptotically median unbiased estimates and for all $\theta \in \Theta$

$$(2.5.1) \quad \overline{\lim}\, P^n\{\theta + t_1\delta_n < T_n < \theta + t_2\delta_n\} \leqslant$$

$$\leqslant F^o(k(t_2);t_2\theta) - (1 - F^o(k(t_1)-;t_1,\theta)) - \gamma_1-\gamma_2$$

where

$$\gamma_i = \frac{[F^1(k\ (t_i);t_i,\theta)-1/2].[F^o(k(t_i);t_i,\theta)-F^o(k(t_i)-;t_i,\theta)]}{[F^1(k(t_i);t_i,\theta) - F^1(k(t_i)-;t_i,\theta)]}; \ i=1,2$$

and $\gamma_i = 0$ if it is of the form 0/0

Proof: The proof runs in the same way as in Pfanzagl (1970) with slight modifications. We have for all sufficiently large n and all $\varepsilon_1 > 0$

$$P^n_{\theta+t_2\delta_n}\{T_n > \theta + t_2\delta_n\} > 1/2 - \varepsilon_1$$

We consider now the test for the hypotheses $\theta$ against the alternative $\theta + t_2\delta_n$ defined by

$$\Phi_n(\underline{x},\varepsilon) = \begin{cases} 1 & \text{if } \ l_n(t_2,\theta) > k + \varepsilon \\ \phi_n & \text{if } \ l_n(t_2,\theta) \in (k-\varepsilon,k+\varepsilon] \\ 0 & \text{if } l_n(t_2,\theta) \leqslant k-\varepsilon \end{cases}$$

with $\phi_n = [F^1_n(k+\varepsilon;t_2,\theta) - 1/2 - \varepsilon_1]/[F^1_n(k_2+\varepsilon;t_2,\theta) - F_n(k_2-\varepsilon;t_2,\theta)]$
clearly if we take $k = k(t_2)$ we have

$$E(\Phi_n(\underline{x},\varepsilon)|H_1)= 1/2 - \varepsilon_1 \text{ for all } n$$

and therefore

$$P^n_{\theta+t_2\delta_n}\{T_n \geqslant \theta + t_2\delta_n\} > E(\Phi_n(\underline{x},\varepsilon)|H_1)$$

Considering now the test $\psi$ for $H_o$ against $H_1$ based on $T_n$ namely $\psi_n(\underline{x}) = 1_{\{T_n\geqslant\theta+t_2\delta_n\}}(\underline{x})$ we obtain in the same way as in the Neyman-Pearson lemma

$$0 > E(\Phi_n(\underline{x},\varepsilon)|H_1) - E(\psi_n(\underline{x})|H_1) \geqslant k.[E(\Phi_n(\underline{x},\varepsilon)|H_o) -$$

$$- E(\psi_n(\underline{x})|H_o)] - \varepsilon.[E(\Phi_n(\underline{x},\varepsilon)|H_o) + E(\psi_n(\underline{x})|H_o)]$$

which gives

$$P_\theta^n\{T_n \geq \theta + t_2\delta_n\} \geq E[\Phi_n(\underline{x},\varepsilon)|H_o]-\varepsilon_2$$

with $\varepsilon_2 \leq 2\varepsilon/k$.

Now observe that

$$E(\Phi_n(\underline{x},\varepsilon)|H_o) = 1-F_n^o(k+\varepsilon;t_2,\theta) + \phi_n.[F_n^o(k_2+\varepsilon;t_2,\theta)-F_n^o(k_2-\varepsilon;t_2,\theta)]$$

and this expression converges for all $|\varepsilon| < \varepsilon_o$. Because of (2.3.2) we obtain

$$\underline{\lim}\ P_\theta^n\{T_n < \theta + t_2\delta_n\} < F^o(k+\varepsilon;t_2,\theta) + \varepsilon_2 - \frac{[F^1(k+\varepsilon;t_2,\theta)-1/2-\varepsilon_1].[F^o(k_2+\varepsilon;t_2,\theta) - F^o(k_2-\varepsilon;t_2,\theta)]}{F^1(k+\varepsilon,t_2,\theta) - F^1(k-\varepsilon,t_2,\theta)}$$

but since $\varepsilon_1$, $\varepsilon_2$ and $\varepsilon$ can be made arbitrarily small we obtain the first part of the bound in (2.4.1). In the same way we get the second bound which gives the desired result.

## 3. Applications in the LAN case

### (3.1) Proposition:

Let $a(t,\theta)$ in (2.1) be of the form $a(t,\theta) = I(\theta).t^2/4$ for an appropriate sequence $\{\delta_n\}$ and for all $t$ and $\theta$ and assume further that for all $t$ and $\theta$

(3.1.1) $\quad \lim\limits_{n\to\infty} n.P_\theta\{f(x,\theta + t\delta_n)/f(x,\theta) < 1-\varepsilon\} = 0$ for all $\varepsilon > 0$

(3.1.2) $\quad \lim\limits_{n\to\infty} n.P_{\theta+t\delta_n}\{f(x;\theta+t\delta_n)/f(x,\theta) > 1+\varepsilon\} = 0$ for all $\varepsilon > 0$

Then for every sequence of estimates $\{T_n\} \in \mathcal{M}$ and for all $t_1,t_2 > 0$ and for all $\theta$

(3.1.3) $\overline{\lim}\ P_\theta^n\{\theta - t_1\delta_n < T_n < \theta + t_2\delta_n\} \leq \phi(t_2.I(\theta)^{1/2}) -$

$- \phi(-t_1.I(\theta)^{1/2})$

Proof: This follows immediately from the fact that under (2.1) (3.1.1) and (3.1.2) is equivalent with contiguity of the sequences $\{P_\theta^n\}$ and $\{P_{\theta+t\delta_n}^n\}$ and asymptotic normality of the loglikelihood-ratios (Osterhoff, Van Zwet (1979)), i.e.

$$\mathcal{L}(\log l_n(t,\theta)|P^n) \to N(-I(\theta).t^2/2, I(\theta).t^2)$$

and

$$\mathcal{L}(\log l_n(t,\theta)|P_{\theta+t\delta_n}^n) \to N(I(\theta)t^2/2, I(\theta).t^2).$$

Thus 2.4 holds and we obtain (2.5.1) with $\gamma_i = 0 \quad i = 1,2$.

Proposition (3.1) contains as a special case the result of Michel (1978), where $\delta_n = n^{-1/2}$ and $I(\theta)$ is the Fisher information but it also covers the so called almost smooth case treated by Ibragimov and Hasminskij (1973), where $\delta_n = [n \log n]^{-1/2}$.

Examples of that kind are the family of triangular distributions with location parameter or the family $f(x,\theta) = C.\exp[-|x-\theta|^{1/2}]$. In all that cases one has a number of estimates which reach the bounds of the covering probabilities e.g. maximum likelihood estimates, maximum probability estimates or Pitman estimates. We will formulate a theorem for the maximum probability estimate which shows that in the LAN case efficiency in the class $\mathcal{M}$ means the same as efficiency in the Wolfowitz class.

(3.2) Theorem:

Assume that the following assumptions hold

(3.2.1) $\lim n\ d^2(\theta+s\delta_n, \theta+t\delta_n) = \frac{1}{4}.|s-t|^2 I(\theta)$ for all $\theta \in \Theta$ and $s,t \in \mathbb{R}$

(3.2.2) $\lim\limits_{n\to\infty} n\ P_{\theta+s\delta_n}\{f(x,\theta+t\delta_n)/f(x,\theta+s\delta_n) < 1-\varepsilon\} = 0$ for all $e>0$

(3.2.3) $\lim\limits_{n\to\infty} n\ P_{\theta+t\delta_n}\{f(x,\theta+t\delta_n)/f(x,\theta+s\delta_n) > 1+\varepsilon\} = 0$ for all $\varepsilon>0$

Then the maximum probability estimate with respect to the interval $(-r,r)$ $M_n(x)$ is strongly asymptotically median unbiased and for $M_n$ we have

$$\text{(3.2.4)} \qquad \lim_{n\to\infty} P_\theta^n\{\theta - t_1\delta_n < M_n(\underline{x}) < \theta + t_2\delta_n\} =$$

$$= \phi(t_2 . I(\theta)^{1/2}) - \phi(-t_1 I(\theta)^{1/2})$$

for all $t_1$, $t_2$ and all $\theta$.

<u>Proof:</u> The result follows from the fact that the distribution of $\delta_n{}^{-1}(M_n-\theta)$ converges continuously in $\theta$ to a normal distribution $N(0,t^2 I(\theta))$. This is proved in the same way as the asymptotic normality of the maximum probability estimate in Grossmann (1979), theorems (3.1) - (3.3) by substituting $\theta_o$ by $\theta_o + s\delta_n$. We therefore only sketch the main steps and refer to the relevant theorems there. First observe that the process

$$\Lambda_n(t) = \log f_n(\underline{x},\theta + t\delta_n)/\log f_n(\underline{x},\theta+s\delta_n)$$

converges to a gaussian process $Z(t)$ with mean

$$\mu(t) = -|t-s|^2 . I(\theta)/2$$

and covariance function

$$r(t_1,t_2) = \frac{1}{2} . I(\theta)(t_1 . t_2 - t_1 . s - t_2 . s + s^2)$$

and thus $Z(t)$ can be represented in the form

$$Z(t) = \sqrt{I(\theta)} . |t-s| . \xi - |t-s|^2$$

where $\xi$ is a standard normal distributed random variable. Next one shows by the same method as in Theorem (3.2) that the finite dimensional distributions of the process

$$L_n(u) = \int_{u-r}^{u+r} [\exp \Lambda_n(t)]dt$$

converges to those of the process

$$L(u) = \int_{u-r}^{u+r} [\exp Z(t)]dt.$$

Now observe that

$$P^n_{\theta+s\delta_n}\{\delta_n^{-1}(M_n-(\theta + s\delta_n)) \leq y\} = P^n_{\theta+s\delta_n}\{M_n \leq \theta+(y-s)\delta_n\} =$$

$$= P^n_{\theta+s\delta_n}\{\sup_{u<y} L_n(u+s) - \sup_{u\geq y} L_n(u+s) \geq 0\}$$

(The equality holds if we take for $M_n$ the infimums of all values where the maximum is obtained.) Now the process $L_n(u)$ has continuous sample pathes and therefore

$$\sup_{u\leq y} L_n(u+s) - \sup_{u>y} L_n(u+s)$$

is continuous functional on the space $C(\mathbb{R})$ and we have

$$P^n_{\theta+s\delta_n}\{\delta_n^{-1}(M_n-\theta + s\delta_n) < y\} \to P\{\sup_{u<y} L(u+s) -$$

$$- \sup_{u\geq y} L(u+s) \geq 0\}$$

But now the process $L(u+s)$ has a unique maximum for

$$u + s = \xi/\sqrt{I(\theta)} + s$$

and this shows that

$$\mathcal{L}(\delta_n^{-1}(M_n-(\theta + s\delta_n))|P^n_{\theta+s\delta_n}) \to N(0,I(\theta)^{-1})$$

and hence we have continuous convergence of the distributions.

## 4. Densities with discontinuities

Ibragimov and Hasminskij (1972) considered families where the densities have a finite number of discontinuities of the first kind and are smooth otherwise. They showed that the limit law of $\mathcal{L}(l_n(t,\theta)|P^n_\theta)$ is the sum of independent poisson distributions which are associated to the points of discontinuity. However in oder to apply our theorem (2.5) we also need the limit law of $\mathcal{L}(l_n(t,\theta)|P^n_{\theta+t\delta_n})$. We will not treat the problem in full generality here but only for the important case where the trace of the measures $P_\theta$ varies in a regular way e.g. like in the case of exponential distributions. We will first state the assumptions we need.

(4.1) Assumptions:

(4.1.1) There exists a function $x(\theta)$ which is differentiable and for all $\theta$ we have $0 < c_1 < x'(\theta) < c_2$.

(4.1.2) For all $\theta$ we have $f(x,\theta) = 0$ if $x < x(\theta)$ and
$$\lim_{x \to x(\theta)+} f(x,\theta) = p(\theta) > 0.$$

(4.1.3) For all $\theta$ and $t$
$$\overline{\lim_{n\to\infty}}\, n\|P_{\theta+t/n} - P_\theta\|_1 =$$
$$= \overline{\lim_{n\to\infty}}\, n.\int|f(x,\theta + t/n) - f(x,\theta)|dx \leq K$$

(4.1.4) For all $\theta$, $t$ and $\delta > 0$
$$\lim_{n\to\infty} n.P_\theta\{f(x,\theta + t/n)/f(x,\theta) < 1-\delta|A_n\} = 0$$
$$\lim_{n\to\infty} n.P_{\theta+t/n}\{f(x,\theta + t/n)/f(x,\theta) > 1+\delta|A_n\} = 0$$
where $A_n = \{x|x > \max(x(\theta),x(\theta + t/n))\}$

(4.1.1) and (4.1.2) are simplifications of the conditions of Ibragimov and Hasminskij (1972) while (4.1.3) and (4.1.4) are the essential of the differentiability conditions stated there. In particular (4.1.1) identifies the sequence $\delta_n$ in condition (2.1) with $\delta_n = 1/n$ as we will see in the following theorem.

(4.2) Theorem:

Let assumption (4.1) hold then

$$\lim_{n\to\infty} n.d^2(\theta,\theta + t/n) = a(t,\theta) = |t|.p(\theta).x'(\theta)$$

Furhermore the limit laws of $\mathcal{L}(l_n(t,\theta)|P_\theta^n)$ and $\mathcal{L}(l_n(t,\theta)|P_{\theta+t/n}^n$ are defined by the following degenerated distributions:

a) for $t > 0$

$$P_o = (1-\exp[-a(t,\theta)]).\delta_{\{o\}} + \exp[-a(t,\theta)].\delta_{\{\exp[a(t,\theta)]\}}$$

$$P_1 = \delta_{\{\exp[a(t,\theta)]\}}$$

b) for $t < 0$

$$P_o = \delta_{\{\exp[-a(t,\theta)]\}}$$

$$P_1 = \exp[-a(t,\theta)].\delta_{\{\exp[-a(t,\theta)]\}} + (1-\exp[-a(t,\theta)]).\delta_{\{\infty\}}$$

where $\delta_{\{x\}}$ denotes the point measure of x.

<u>Proof:</u> We keep some $\theta$ and $t > 0$ fixed and because of (4.1.1) we have $A_n = \{x|x > x(\theta + t/n)\}$ for all n. Furthermore

$$P_\theta(A_n) = 1-\int_{x(\theta)}^{x(\theta+t/n)} f(x,\theta)dx = 1-\frac{t}{n}.x'(\theta).p(\theta) + o(1/n)$$

because of (4.1.1) and (4.1.2). Define now

$$\Lambda_n(x) = \log[f(x,\theta + t/n)/f(x,\theta)].1_{A_n}(x)$$

where $1_{A_n}(x)$ is the indicator function of $A_n$. We will first show that

$$E[\Lambda_n(x)] = \frac{t}{n}.x'(\theta).p(\theta) + o(1/n).$$

Consider therefore

$$E_\theta[\Lambda_n(x)] = \int_{A_n\cap B} \Lambda(x)f(x,\theta)dx + \int_{A_n\cap B_1} \Lambda(x)f(x,\theta)dx +$$

$$+ \int_{A_n\cap B_2} \Lambda(x)f(x,\theta)dx = I_1 + I_2 + I_3$$

where

$$B_1 = \{\Lambda(x) < -\delta\},\ B_2 = \{\Lambda(x) > \delta\}\quad B = \bar{B}_1\cap\bar{B}_2$$

For the second integral we have

$$|I_2| < |\int_{A_n \cap B_1} [f(x,\theta + t/n) - f(x,\theta)]dx| \leq$$

$$\leq P_{\theta+t/n}(B_1 \cap A_n) + P_\theta(B_1 \cap A_n)$$

Since $B_1 \subset \{f(x,\theta + t/n)/f(x,\theta) < 1-\delta_1\}$ for some $\delta_1$ we obtain from (4.1.4) for sufficiently large n

$$P_\theta(B_1 \cap A_n) < \varepsilon/n, \; P_{\theta+t/n}(B_1 \cap A_n) < \varepsilon/n$$

hence $|I_2| \leq 2\varepsilon/n$.
In the same way one concludes $|I_3| \leq 2\varepsilon/n$.
To compute $I_1$ we use the inequality $|\log(1+y)-y| < \delta.|y|$ for $|y| < \delta$ and $\delta$ sufficiently small. This yields to

$$I_1 = \int_{A \cap B} [f(x,\theta + t/n) - f(x,\theta)]dx + R$$

with

$$|R| \leq \delta. \int_{A_n \cap B} |f(x,\theta + t/n) - f(x,\theta)|dx \leq$$

$$\leq \delta.\int|f(x,\theta + t/n) - f(x,\theta)|dx \leq \delta.k/n$$

because of (4.1.3).
Now

$$\int_{A_n} [f(x,\theta + t/n) - f(x,\theta)]dx = 1- \int_{x(\theta+t/n)}^{\infty} f(x,\theta)dx =$$

$$\frac{t}{n}.x'(\tilde{\theta}).f(\tilde{x},\theta)$$

and because the set $\bar{B}$ has $P_\theta$ probability less then $\varepsilon/n$ we get

$$E_\theta[\Lambda(x)1_A(x)] = \frac{t}{n}.x'(\theta).p(\theta) + o(1/n)$$

By the same arguments one obtains

$$E_{\theta+t/n}[\Lambda_n(x)] = -E_{\theta+t/n}[\log[f(x,\theta)/f(x,\theta + t/n)].$$

$$.1_{A_n}(x)] = t.x'(\theta).p(\theta)/n + o(1/n)$$

Now

$$d^2(\theta,\theta + t/n) = 2.[1-\int_{A_n} f(x,\theta + t/n)^{1/2}.f(x,\theta)^{1/2}dx] =$$

$$= 2.\int_{A_n} \frac{f(x,\theta+t/n)^{1/2}-f(x,\theta)^{1/2}}{f(x,\theta + t/n)^{1/2}} . f(x,\theta + t/n)dx$$

By the same methods as before we split the integral into the three parts on $B$, $B_1$ and $B_2$ and the integrals over $B_1$ and $B_2$ are smaller then $2\varepsilon/n$. For the integral over $B \cap A_n$ we use the expansion

$$- \frac{f(X,\theta+t/n)^{1/2}-f(x,\theta)^{1/2}}{f(x,\theta+t/n)^{1/2}} =$$

$$= \frac{1}{2} \log[f(x,\theta)/f(x,\theta + t/n)] + R(x)$$

with

$$\int_{A_n \cap B} |R(x)|dx \leqslant \delta \int_{A_n \cap B} |f(x,\theta + t/n)^{1/2} -$$

$$- f(x,\theta)^{1/2}|f(x,\theta + t/n)^{1/2}dx \leqslant$$

$$\leqslant \delta \int_{A_n \cap B} |f(x,\theta + t/n) - f(x,\theta)|dx \leqslant \delta.k/n$$

because of $|\sqrt{x}.\sqrt{y} - y| < |x-y|$ if $1-\delta < \frac{x}{y} < 1+\delta \quad x > 0,\ y > 0$ and this gives

$$n.d^2(\theta,\theta + t/n) = t.x'(\theta).p(\theta) + o(1)$$

Next we show that

$$E_\theta[\Lambda_n^2(x)] = o(1/n) \text{ and } E_{\theta+t/n}(\Lambda_n(x))^2 = o(1/n)$$

To see this observe first that

$$E_\theta[\Lambda_n^2(x)] = 4.E_\theta\{[\log(1+\frac{f(x,\theta+t/n)^{1/2}-f(x,\theta)^{1/2}}{f(x,\theta)^{1/2}})]^2.1_{A_n}(x)\}$$

and

$$|\log(1+y^2) - y^2)| < 2\delta.|y| \text{ for } |y| < \delta < 1/2$$

Now splitting up again into the sets B, $B_1$, $B_2$ we get

$$E_\theta[\Lambda_n(x)^2] = 4. \int_{A_n \cap B} [f(x,\theta + t/n)^{1/2} -$$

$$- f(x,\theta)^{1/2}]^2 dx + o(1/n) =$$

$$= 4.[d^2(\theta,\theta + t/n) - P_\theta(\bar{A}_n)] + o(1/n)$$

and since $d^2(\theta,\theta + t/n) = t.x'(\theta).p(\theta)/n+o(1/n) = P_\theta(\bar{A}_n) + o(1/n)$ we obtain the desired result.

But now follows easily the conclusions of the theorem. By the weak law of large numbers (e.g. Petrov (1975) p. 258 TH.1) we get

$$\lim_{n\to\infty} \sum_{i=1}^{n} \log[f(x_i,\theta + t/n)/f(x_i,\theta)].1_{A_n}(x_i) =$$

$$= t.x'(\theta).p(\theta)$$

and since

$$P_\theta^n\{\min(x_i) < x(\theta + t/n)\} = 1-P_\theta\{x_i \leqslant x(\theta + t/n)\}^n =$$

$$= 1-(1- \frac{t.x(\theta).p(\theta)+o(1)}{n})^n$$

and

$$P_{\theta+t/n}^n\{\min(x_i) < x(\theta + t/n)\} = 0$$

we obtain the desired results for $t > 0$. In the case $t < 0$ observe that $A_n = \{x | x > x(\theta)\}$ for all n and $l_n(t,\theta) = \infty$ for $\min(x_i) < x(\theta)$ which completes now in the same way as above the proof of the theorem.

(4.3) Corollary:

Let assumptions (4.1) hold and let $t_i$ $i = 1,2$ be positive with $t_1 < \log 2/p(\theta).x'(\theta)$. Then for every sequence $\{T_n\} \in \mathcal{M}$ we have

(4.3.1) $$\overline{\lim}\, P^n\{\theta - t_1/n < T_n < \theta + t_1/n\} \leqslant$$

$$\leqslant [\exp[p(\theta).x'(\theta).t_1]-\exp[-p(\theta).x'(\theta).t_2]]/2$$

Proof: Because of (4.2) we have in (2.4.2)

$$k(t_2) = \exp[p(\theta).x'(\theta).t_2]$$
$$k(-t_1) = \exp[-p(\theta).x'(\theta).t_1]$$

and now direct computation of the bounds of theorem (2.5) gives the result.

We will next show that in the case of location families the bound can be obtained.

(4.4) Proposition:

Let assumptions (4.1) hold and let furthermore $\{P_\theta\}$ be a location family with $f(x,\theta) = f(x-\theta)$ then the sequence of estimates $\{\hat{T}_n\}$ defined by $T_n(\underline{x}) = \min(x_i) - \ln 2/n.p$ with $p = \lim_{x \to 0+} f(x)$ is strongly asymptotically median unbiased and is an efficient estimate for the class $\mathcal{M}$.

Proof: First observe that $x(\theta) = \theta$ and $p(\theta) = p$.
Now

$$P^n_{\theta+t/n}\{\min(x_i) - \ln 2/n.p \geqslant \theta + t/n\} =$$

$$= P_0\{ x \geqslant \ln 2/n.p\}^n = [1 - \ln 2/n]^n$$

and therefore $\{\hat{T}_n\} \in \mathcal{M}$
In the same way one calculates directly for $t_1 < \ln 2/p$

$$\lim_{n\to\infty} P^n_\theta \{\theta - t_1/n < \min(x_i) - \ln 2/p.n < \theta + t_2/n\} =$$

$$= [\exp(p.t_1) - \exp(-p.t_2)]/2$$

which shows all.

(4.5) Remarks:

In the location case of proposition (4.4) the maximum probability estimate for a symmetric interval $(-r,r)$ is given by

$M_n(\underline{x}) = \min(x_i) - r/n$. If we compute the covering probability for $M_n(x)$ we obtain

$$\lim P_\theta^n\{\theta - r/n \leqslant \min(x_i) - r/n \leqslant \theta + r/n\} =$$

$$1-\exp[-2r.p]$$

and hence the bounds for the maximum probability estimate are only equal to the bounds of the efficient estimate in $\mathcal{M}$ if we take $t_1 = t_2 = \log 2/p$ and also $r = \log 2/p$ which is also the only case where the maximum probability estimate is also in the class $\mathcal{M}$. In all the other cases if we take $t_1 = t_2 = r$ and less then $\log 2/p$ the maximum probability estimate $M_n(x)$ has a greater covering probability, whereas if we use for comparism an interval $(\ln 2, r)$ with $r > \log 2/p$ the most efficient strongly asymptotically median unbiased estimate is more efficient than the maximum probability estimate. This shows that the two concepts of efficiency are in general not comparable.
Similar results can be obtained for densities where $f(x,\theta) = 0$ for $x > x(\theta)$ and the other assumptions in (4.1) appropriate modified or for the case $f(x,\theta) = 0$ for $x < x_1(\theta)$ or $x > x_1(\theta)$. However in that case it is in general not possible to give efficient estimates even in the location case. Akahira and Takeuchi (1979) showed that for truncated normal families an efficient estimates exist in $\mathcal{M}$. But the example stated there cannot be generalized to the family of uniform distribution $f(x,\theta) = 1_{[\theta,\theta+1]}$.

References

Akahira M., Takeuchi K. (1979). Asymptotic Efficiency of Estimates. (unpublished manuscript)

Grossmann W. (1979). Einige Bemerkungen zur Theorie der Maximum Probability Schätzer. Metrika 26, 129 - 137.

Ibragimov I.A, Hasmiskij R.Z. (1972). Asymptotic behavior of statistical estimates for samples with a discontinuous density. Math. USSR Sbornik 16, 573-606.

Ibragimov I.A., Hasminskij R.Z. (1973). Asymptotic analysis of statistical estimators for the "almost smooth" case. Theor. Prob. Appl. 18, 241-252.

Michel R. (1978). On the asymptotic efficiency of strongly asymptotically median unbiased estimators. Ann. Stat. 6, 920 - 922.

Oosterhoff J., Van Zwet W.R.(1979). A note on contiguity and Hellinger distance. In Contributions in Statistics 157 - 166, Reidel, London.

Petrov V.V. (1975). Sums of independent random variables. Springer, Berlin.

Pfanzagl J. (1970). On the asymptotic efficiency of median unbiased estimates. Ann. Math. Statist 41, 1500 - 1509.

Wolfowitz J. (1974). Maximum probability estimators and related topics. Springer, Lecture Notes in Mathematics 424.

LINEAR FORMS IN RANDOM VARIABLES DEFINED ON A HOMOGENEOUS MARKOV CHAIN

Béla Gyires

Kossuth L. University of Debrecen ( Hungary )

## 1. Introduction

The following classical result of G. Darmois and V.P. Skitovich ( [4] , 97 ; [6] 75 ) is well-known.

Let $a_j, b_j$ ( $j = 1, \ldots, n$ ) be real numbers, and let $\xi_1, \ldots, \xi_n$ be independently, but not necessarily identically, distributed random variables. Suppose that the linear forms $a_1 \xi_1 + \cdots + a_n \xi_n$ and $b_1 \xi_1 + \cdots + b_n \xi_n$ are independently distributed. Then each random variable $\xi_j$, which has non-zero coefficients in both forms, is normally distributed.

This paper deals with the same problem, but in the case, if the random variables $\xi_1, \ldots, \xi_n$ are not independent, however these are defined on a homogeneous Markov chain. This dependence idea was introduced by the author, who used this to give a generalization of the sufficiency half of the Lindeberg-Feller central limit theorem for sums of a sequence of random variables ( [3] ), and to give a new characterization of the Normal Law ( [1] ).

Besides the introduction this paper consists of three other chapters. The second one treats the definitions and lemmas, wich are necessary in the following. The third chapter contains the extensions of the over mentioned theorem of Darmois and Skitovich.

We say that the matrices

$$A_k(x) = (a_{hj}^{(k)}(x)) \quad (k = 1, \ldots, 2n;\ x \in R_1)$$

formed by the elements

$$a_{hj}^{(k)}(x) = P(\xi_k < x,\ \eta_k = j \mid \eta_{k-1} = h),$$

and the matrices

$$\varphi_k(t) = \int_{-\infty}^{\infty} e^{itx}\, dA_k(x) \quad (k = 1, \ldots, 2n;\ t \in R_1) \tag{2,4}$$

are matrix values distribution functions and matrix values characteristic functions defined on the homogeneous Markov chain $\{A\}$ respectively.

In the next we are using the following lemmas.

Lemma 2.1.

Suppose that the random variables $\xi_1, \ldots, \xi_{2n}$ are defined on the homogeneous Markov chain $\{A\}$. Then

$$a_{hj}^{(k)}(x) = a_{hj}\, F_{hj}^{(k)}(x),$$

where $F_{hj}^{(k)}(x)$ is the distribution function of $\vartheta_{hj}^{(k)}$.

Proof.

See [1], Lemma 2.1.

Lemma 2.2.

Suppose that the random variables $\xi_1, \ldots, \xi_{2n}$ are defined on the homogeneous Markov chain $\{A\}$. If

$$1 \leqq i_1 < \cdots < i_n \leqq 2n, \tag{2,5}$$

then

$$\left(P(\xi_{i_1} < x_1, \ldots, \xi_{i_n} < x_n,\ \eta_{2n} = j \mid \eta_0 = h)\right)_{h,j=1}^{r} =$$

## 2. Preliminaires

In the following chapters we are talking about $p \times p$ matrices. Denote $E$ the unit matrix.

Under the derivative and integration of a matrix values function we mean the matrix formed from the derivatives and integrations of the elements of the matrix values function respectively.

Suppose that

$$(2,1) \qquad \vartheta_{hj}^{(k)} \qquad (h, j = 1, \ldots, p;\ k = 1, 2, \ldots)$$

a re independently distributed random variables.

Let the random variables

$$(2,2) \qquad \{\eta_j\}_{j=0}^{\infty}$$

form a homogeneous Markov chain with states $1, \ldots, p$ and with transition matrix

$$(2,3) \qquad A = (a_{hj}).$$

Denote $\{A\}$ this homogeneous Markov chain.

Let (2,1) and (2,2) be independent random variables to the effect that all finite dimension vectors with components (2,1) are independent of all finite dimension vectors with components (2,2).

Definition 2.1.

We say that the random variables $\xi_1, \ldots, \xi_{2n}$ are defined on the homogeneous Markov chain $\{A\}$, if

$$\xi_k = \vartheta_{\eta_{k-1}\eta_k}^{(k)} \quad (k = 1, \ldots, 2n),$$

i. e. $\xi_k = \vartheta_{hj}^{(k)}$, if $\eta_{k-1} = h$, $\eta_k = j$. •

$$= A^{i_1-1} A_{i_1}(x_1) A^{i_2-i_1-1} A_{i_2}(x_2) \ldots A^{i_m-i_{m-1}-1} A_{i_m}(x_m) A^{2m-i_m}$$

Proof.

See [1] , Lemma 2.2.

Denote $\mathcal{P}_q$ the set of the matrix values polynomials of degree $q$ , in which the coefficient of the $q$ -th exponent has the form $\lambda U$ , where $\lambda > 0$ is a real number and $U$ is an unitary matrix.

Lemma 2.3.

Let the stochastic matrix (2,3) be given and let $\operatorname{tr} A \leqq \sqrt{p}$. Moreover let $Q(t) \in \mathcal{P}_q$ . Then $\varphi(t) = \exp\{Q(t)\}$ cannot be a matrix values characteristic function defined on the homogeneous Markov chain $\{A\}$ for $q \geqq 3$ .

Proof.

See [2] , Corollary 3.1.

Lemma 2.4.

Let the elements of the stochastic matrix (2,3) be positive numbers, and let $\operatorname{tr} A \leqq \sqrt{p}$ . Let the coefficients of $Q(t) \in \mathcal{P}_q$ be commutable with one another. Then $\varphi(t) = \exp\{Q(t)\}$ is a matrix values characteristic function defined on the homogeneous Markov chain $\{A\}$ , if and only if $Q(t) = A \exp\{i\gamma t - \lambda t^2\}$, where now $\lambda \geqq 0$ and $\gamma$ is a real number.

Proof.

See [2] , Theorem 3.3.

Let the real numbers $a_j \neq 0$ , $b_j \neq 0$ ( $j = 1, \ldots, m$ ) be given. The elements $1, \ldots, 2m$ , which are different to (2,5) let us denote by $k_1 < \cdots < k_m$. We construct now the linear forms

$$L_1(i_1,\dots,i_m) = a_1\xi_{i_1} + \dots + a_m\xi_{i_m},$$

(2,6)

$$L_2(i_1,\dots,i_m) = b_1\xi_{i_1} + \dots + b_m\xi_{i_m},$$

where $\xi_1,\dots,\xi_{2n}$ are random variables defined on the homogeneous Markov chain $\{A\}$ .

Definition 2.2.

We say that the random variable $L_1(i_1,\dots,i_m)$ is independent to the random variable $L_2(i_1,\dots,i_m)$ on the homogeneous Markov chain $\{A\}$ , if

$$\left(E\left(\exp\{itL_1(i_1,\dots,i_m)+iuL_2(i_1,\dots,i_m)\},\eta_{2n}=j\,|\,\eta_0=h\right)\right)_{h,j=}^{r} =$$

$$= \prod_{k=1}^{2n}\varphi_k(c_k v),\quad t\in R_1,\ u\in R_1,$$

where

$$(2,7)\qquad \left.\begin{aligned} c_{i_l} &= a_{i_l}, \quad v=t \\ c_{k_l} &= b_{i_l}, \quad v=u \end{aligned}\right\}(l=1,\dots,m).$$

Definition 2.3.

The forms $L_1$ and $L_2$ are completely independent on the homogeneous Markov chain $\{A\}$ , if the random variable $L_1(i_1,\dots,i_m)$ is independent to the random variable $L_2(i_1,\dots,i_m)$ on the homogeneous Markov chain $A$ for all $i_1,\dots,i_m$ satisfying inequality (2,5).

Since

$$t L_1(i_1,\dots,i_m) + u L_2(i_1,\dots,i_m) =$$

$$= (a_1 t + b_1 u)\xi_{i_1} + \dots + (a_m t + b_m u)\xi_{i_m},$$

on the basis of Lemma 2.2. we get the following result.

Theorem 2.1.

Let the random variables $\xi_1,\dots,\xi_{2m}$ be defined on the homogeneous Markov chain $\{A\}$. Then the random variable $L_1(i_1,\dots,i_m)$ is independent to the random variable $L_2(i_1,\dots,i_m)$ on the homogeneous Markov chain $\{A\}$, if and only if the matrix values functional equation

$$A^{i_1-1}\varphi_{i_1}(a_1 t + b_1 u) A^{i_2-i_1-1}\varphi_{i_2}(a_2 t + b_2 u)\dots$$

(2,8) $$\dots A^{i_m-i_{m-1}-1}\varphi_{i_m}(a_m t + b_m u) A^{2m-i_m} =$$

$$= \prod_{k=1}^{2m} \varphi_k(C_k \vartheta), \quad t \in R_1, \ u \in R_1$$

is satisfied by the matrix values characteristic functions $\varphi_{i_\ell}(t)$ ( $\ell = 1,\dots,m$ ), where the quantities $C_k \vartheta$ are defined by (2,7).

Corollary 2.1.

Under the conditions of Theorem 2.1. the forms $L_1$ and $L_2$ are completely independent on the homogeneous Markov chain $\{A\}$, if and only if the matrix values functional equation (2,8) is satisfied by the matrix values characteristic functions (2,4) for all $i_1,\dots,i_m$ satisfying the inequality (2,5).

If the homogeneous Markov chain $\{A\}$ and the real numbers $a_j$ , $b_j$ ( $j=1,\dots,m$ ) are given, then we can take up the following two questions.

a./ Under what kind of matrix values characteristic functions will be the random variable $L_1(i_1,\dots,i_m)$ independent to the random variable $L_2(i_1,\dots,i_m)$ on the homogeneous Markov chain $\{A\}$ ?

b./ Under what kind of matrix values characteristic functions will be the forms $L_1$ and $L_2$ completely independent on the homogeneous Markov chain $\{A\}$ ?

If we want to give an answer to these questions, it is necessary to solve either the matrix values functional equation (2,8) over fixed $i_1,\dots,i_m$ , or the matrix values functional equation system (2,8) over all $i_1,\dots,i_m$ satisfying inequality (2,5) by matrix values characteristic functions. It seems, this task is not very simple.

To simplify this task, in the following we consider only the case, if the commutability conditions

$$(2,9) \qquad \varphi_k(t)\varphi_l(u)=\varphi_l(u)\varphi_k(t)\ ,\ k\neq l,\ t\in R_1,\ u\in R_1$$

are satisfied by the matrix values characteristic functions (2,4). In this case we get the matrix values functional equation

$$(2,10) \qquad A^m\prod_{k=1}^{m}\varphi_{i_k}(a_kt+b_ku)=\prod_{k=1}^{m}\varphi_{i_k}(a_kt)\,\varphi(b_ku)$$

$$(1\leqq i_1<\dots<i_m\leqq 2m,\ t\in R_1,\ u\in R_1)$$

from (2,8).

In chapter 3 we deal with the solution of functional

equation (2,10).

## 3. The generalization of the theorem of Darmois and Skitovich

Theorem 3.1.

Suppose that the commutatbility conditions (2,9) are satisfied by the matrix values characteristic functions (2,4) of the random variables defined on the homogeneous Markov chain $\{A\}$, and let $\operatorname{Det} A \neq 0$. Moreover let the forms (2,6) be completely independent of the homogeneous Markov chain $\{A\}$. Then

$$(3,1)\qquad \varphi_k(t) = \exp\{Q_k(t)\} \qquad (k = 1,\dots,2n),$$

where $Q_k(t)$ is a quadratic matrix values polynomial.

Proof.

Since $A$ is a regular matrix, the matrices of the both sides of (2,9) are regular too in a neighboushood $|t| < \delta$, $|u| < \delta$, $\delta > 0$. Let

$$\log\{A\,\varphi_k(t)\} = \Phi_k(t) \quad (k = 1,\dots,2n)$$

in this neighbourhood. Then the expression

$$(3,2)\qquad \sum_{j=1}^{m} \Phi_{i_j}(a_j t + b_j u) = A_{i_1\dots i_m}(t) + B_{i_1\dots i_m}(u)$$

take the place of the functional equation (2,10), where

$$A_{i_1\dots i_m}(t) = \sum_{j=1}^{m} \log \varphi_{i_j}(a_j t),$$

$$B_{i_1\dots i_m}(u) = \sum_{j=1}^{m} \log \varphi_{i_j}(b_j u).$$

Applying a similar procedure in (3,2), which was used in the case of $p = 1$ by E. Lukács and R.G. Laha in their book [6] ( 76-77 ), we get in the neighbourhood at issue

$$A\ \varphi_k(t) = \exp\{P_k(t)\} \quad (k = 1, \ldots, 2n), \tag{3,3}$$

where $P_k(t)$ is a matrix values polynomial of degree not exceedig $n$ . Let

$$P_k(t) = Q_k(t) + R_k(t) \quad (k = 1, \ldots, 2n),$$

where $Q_k(t)$ is a quadratic matrix values polynomial and

$$R_k(t) = \sum_{j=3}^{n} a_j^{(k)} t^j .$$

In order to show that the degree of $P_k(t)$ is not exceeding two, applying a similar procedure, which was used by Yu. V. Linnik in his book [4] ( 98-99 ) in the case of $p = 1$ , we get that

$$\sum_{j=1}^{n} \beta_j^2\ \Phi_{i_j}(t) = P_{i_1 \ldots i_n}(t) \tag{3,4}$$

$$(1 \leqq i_1 < \cdots < i_n \leqq 2n)$$

holds in the neighbourhood at issue, where $P_{i_1 \ldots i_n}(t)$ is a quadratic matrix values polynomial, and

$$\beta_j = \frac{b_j}{a_j} \quad (j = 1, \ldots, n).$$

Let $n < k \leqq 2n$ . Then we get

$$\Phi_k(t) = \Phi_n(t) + \frac{1}{\beta_n^2}\left[P_{1 \ldots n-1\,k}(t) - P_{1 \ldots n}(t)\right]$$

on the basis of (3,4), i. e.

$$Q_k(t) = Q_n(t) + \frac{1}{\beta_n^2}\left[P_{1 \ldots n-1\,k}(t) - P_{1 \ldots n}(t)\right],$$

$$R_k(t) = R_m(t) \quad (k = m, m+1, \ldots, 2m),$$

Using the last formula, we get again on the basis of (3,4), that

$$\sum_{j=1}^{m} \beta_j^2 \Phi_{m+j}(t) = \sum_{j=1}^{m} \beta_j^2 Q_{m+j}(t) +$$

$$+(\beta_1^2 + \cdots + \beta_m^2) R_m(t) = P_{m+1 \ldots 2m}(t).$$

Thus

$$R_k(t) \equiv 0 \qquad (k = m, m+1, \ldots, 2m)$$

i. e. the statement of our theorem holds in a neighbourhood of the origin, if $k = m, m+1, \ldots, 2m$.

Let now $1 \leq k < m$. Then we obtain on the basis of (2,4) too that

$$\beta_1^2 \Phi_k(t) + \sum_{j=2}^{m} \beta_j^2 \Phi_{m+j}(t) = P_{k\, m+2 \ldots 2m}(t),$$

where $\Phi_{m+j}(t)$ ($j = 2, \ldots, m$), and $P_{k\, m+2 \ldots 2m}(t)$ are quadratic matrix values polynomials. Thus $\Phi_k(t)$ ($k = 1, \ldots, m-1$) is a quadratic matrix values polynomial too.

We obtained now that (3,1) holds in a neighbourhood of the origin. Since $\varphi_k(t)$ is analytic in this neighbourhood, therefore (3,1) holds on the whole real line too ( [5] , Theorem 7.1.1.), and this completes the proof of the Theorem 3.1.

In the next we take only the matrix values polynomials into consideration, which are elements of the set $\mathcal{P}_m$ ($m = 1, 2, \ldots$).

<u>Theorem 3.2.</u>

Let the elements of the regular stochastic matrix $A$

be positive numbers, and let $\operatorname{tr} A \leqq \sqrt{p}$ . Suppose that the commutability conditions (2,9) are satisfied by the matrix values characteristic functions of the random variables defined on the homogeneous Markov chain $\{A\}$ for $k, \ell = i_1, \ldots, i_n$ . Moreover let the random variable $L_1(i_1, \ldots, i_n)$ be independent to the random variable $L_2(i_1, \ldots, i_n)$ on the homogeneous Markov chain $\{A\}$ . Then

$$\varphi_k(t) = A \exp\{i \gamma_k t - \lambda_k t^2\}, \tag{3,5}$$

where $\lambda_k \geqq 0$ and $\gamma_k$ are real numbers.

Proof.

Because (3,3) holds now in a neighbourhood of the origin in the case if $P_k(t) \in \mathcal{P}_m$ for $k = i_1, \ldots, i_n$ , we obtain as a consequence of Lemma 2.3., that $P_k(t) \in \mathcal{P}_2$ . Moreover since the conditions of the Lemma 2.4. are satisfied, (3,5) holds in a neighbourhood of the origin and therefore ( [5] , Theorem 7.1.1.) on the whole real line too.

By substitutions $p = 1$ we can get the theorem of Darmois and Skitovich ( [6] , 75 ) from Theorem 3.2.

## References

[1] Gyires, B. Constant regression of quadratic statistics on the sum of random variables defined on a Markov chain. Festschrift volume dedicated to E. Lukács. 1981.

[2] Gyires, B. On an extension of the Marcinkiewicz theorem. J. Multivar. Anal.

[3] Gyires, B. Eine Verallgemeinerung des zentralen Grenzwertsatzes. Acta Math. Acad. Sci. Hung., XIII (1956), 69-80.

[4] Linnik, Yu. V. Decomposition of probability distributions. Oliver and Boyd Ltd, Edinburgh and London, 1964.

[5] Lukács, E. Characteristic functions. Griffin, London, 1960.

[6] Lukács, E. a nd Laha, R. G. Applications of characteristic functions. Hafner Publ. Co., New York, 1961.

Parallel processing of random sequences with priority

By

A. Iványi and I. Kátai

Department of Numerical Mathematics and Computer Sciences,

Eötvös Loránd University, Budapest

Dedicated to B. Gyires on his 70th anniversary

§1. Introduction and main results. We continue our work [1]. Let $\mathcal{A}_N$ denote the set $\{1,2,\ldots,N\}$ and

$$
\begin{array}{l}
f_1^{(1)},\ f_2^{(1)},\ldots \\
f_1^{(2)},\ f_2^{(2)},\ldots \\
\vdots \\
f_1^{(r)},\ f_2^{(r)},\ldots
\end{array}
\qquad /1.1/
$$

an infinite sequence of infinite sequences, the elements of which belong to $\mathcal{A}_N$. We process the elements in the sequences according to the following rules:

1. Processing proceeds in the points of time $i=1,2,\ldots$ . Let $i = 1$.

2. Let $k_1$ denote the largest positive integer for which the elements $f_1^{(1)},\ldots,f_{k_1}^{(1)}$ are distinct. Assuming that $k_1$, $\ldots$, $k_{\ell-1}$ have been defined, let $k_\ell$ denote the largest nonnegative integer for which $f_1^{(1)},\ldots,f_{k_t}^{(1)}$ are distinct and

$$/1.2/ \qquad \bigcup_{j=1}^{l-1} \left\{f_1^{(j)},\dots,f_{k_j}^{(j)}\right\} \cap \left\{f_1^{(l)},\dots, f_k^{(l)}\right\} = \emptyset.$$

3. In the i-th point of time we process the first $k_t$ elements of the t-th /t=1,.../ sequence. We omit the processed elements from the sequences, and reduce the lower index of the remaining elements by $k_t$ in the t'th sequence.

4. We add 1 to i and continue the processing from the rule 2.

Let us suppose that $\xi_i^{(j)}$ /i=1,2,...; j=1,2,.../ are independent random variables with the distribution

$$/1.3/ \qquad P\left(\xi_i^{(j)} = k\right) = \frac{1}{N}, \qquad k \in \mathcal{A}_N.$$

Consider the array

$$/1.4/ \qquad \begin{array}{ll} \xi_1^{(1)}, & \xi_2^{(1)},\dots \\ \vdots & \\ \xi_1^{(r)}, & \xi_2^{(r)},\dots \end{array}$$

Let $p(k_1,k_2,\dots,k_r)$ be the probability of the event that in the first point of time for every $t=1,\dots,r$ the first $k_t$ elements of the t'th sequence are processed. Let $s_1=k_1$, $s_j=k_1+k_2+\dots+k_j$ $(j \geq 2)$. As it is obvious

/1.5/ $$p(k_1,\dots,k_r) = \gamma(s_r-1,N)\,\frac{s_1}{N}\cdots\frac{s_r}{N}\,,$$

where

/1.6/ $$\gamma(t,N) = \prod_{\nu=1}^{t}\left(1-\frac{\nu}{N}\right).$$

Let $\rho$ be the least integer $h$ for which $k_1+\dots+k_h=N$. We write $\rho=\infty$, if $k_1+\dots+k_r<N$ for every $r$.

We can see easily that

/1.7/ $$P(\rho=\infty) = 0.$$

Indeed, if $\rho=\infty$, then there exists a $\gamma\in\mathcal{A}_N$ which does not occur in the sequence $\xi_1^{(1)}, \xi_1^{(2)},\dots$ . Since

/1.8/ $$P\left(\xi_1^{(1)}\neq\nu,\dots,\xi_1^{(h)}\neq\gamma\right)=\left(1-\frac{1}{N}\right)^h\to 0 \qquad (h\to\infty),$$

therefore /1.7/ holds.

For the real numbers $p\in(0,1)$ and $q=1-p$ let $\theta(p)$ be a random variable defined by

/1.9/ $$P(\theta(p)=\ell) = p^\ell q \qquad (\ell=0,1,2,\dots).$$

We shall prove the following

THEOREM 1. The random variable $\rho - 1$ can be written as the sum of the independent ramdom variables $\theta\left(\frac{1}{N}\right), \theta\left(\frac{2}{N}\right), \ldots, \theta\left(\frac{N-1}{N}\right)$, i.e.

/1.10/ $$\rho - 1 = \theta\left(\frac{1}{N}\right) + \theta\left(\frac{2}{N}\right) + \ldots + \theta\left(\frac{N-1}{N}\right).$$

Consequently, by

/1.11/ $$\eta_N = \frac{\rho}{N} - \log N,$$

we have

/1.12/ $$\lim_{N \to \infty} P\left(\eta_N < x\right) = e^{-e^{-x+1}}$$

For an integer $M \in [1, N]$ let $\Lambda(M) = \Lambda_N(M)$ be the least $r$ for which $s_r \geq M$. Let $\phi(x) = \frac{1}{\sqrt{2\pi}} \int_{-\infty}^{x} \exp\left(-\frac{u^2}{2}\right) du.$

THEOREM 2. For any fixed $\alpha \in (0,1)$, with $M = [\alpha N]$ we have

/1.13/ $$\lim_{N \to \infty} P\left(\frac{\Lambda_N M - \left(-\alpha - \log(1-\alpha)\right) N}{\left(\log(1-\alpha) + \frac{\alpha}{1-\alpha}\right) \sqrt{N}} < x\right) = \phi(x).$$

THEOREM 3. Let $M_N$ be such a sequence of integers that $M_N / N \to 0$ and $M_N / \sqrt{N} \to \infty$. Then

$$/1.14/ \quad \lim_{N\to\infty} P\left( \frac{\Lambda_N(M_N) - \frac{1}{2}\,\frac{M_N^{\,2}}{N}}{\sqrt{\frac{1}{2}\cdot\frac{M_N^{\,2}}{N}}} < x \right) = \phi(x).$$

Theorem 3 can be proved on a similar way as Theorem 2, so we shall omit the proof of Theorem 3.

THEOREM 4. Let $A>0$ be a positive constant, $M=M_N$ be such a sequence that $M_N/\sqrt{N}\to A$. Then

$$/1.15/ \quad \lim_{N\to\infty} P\left(\Lambda_N(M_N) = r-1\right) = \exp\left(-\frac{A^2}{2}\right)\cdot\left(\frac{A^2}{2}\right)^{r-1}\frac{1}{(r-1)!}$$

$$(r=1,2,\ldots).$$

§2. Proof of Theorem 1. We start with the formula

$$/2.1/ \quad P(\rho = r) = \sum_{k_1,\ldots,k_r}^{*} p(k_1,\ldots,k_r), \qquad (s_r=k_1+\ldots+k_r\ !)$$

where the star denotes that we sum up for those $k_1,\ldots,k_r$ for which $s_{r-1}\le N-1$ and $s_r=N$ hold. So

$$/2.2/ \quad P(\rho=r) = \prod_{t=1}^{N-1}\left(1-\frac{t}{N}\right) \sum_{1\le s_1\le s_2\ \ldots\le s_{r-1}\le N-1} \frac{s_1\cdots\,.s_{r-1}}{N^{r-1}},$$

where we sum for every integer $s_1,\ldots,s_{r-1}$ indicated in the right hand side. Let $\nu_j$ denote the number of occurences of j in the sequence $s_1,\ldots,s_{r-1}$. Then an appropriate product $\frac{s_1\cdot\ldots\cdot s_{r-1}}{N^{r-1}}$ can be written as

$$/2.3/ \qquad \prod_{j=1}^{N-1}\left(\frac{j}{N}\right)^{\nu_j}, \quad \nu_1+\ldots+\nu_{N-1}=r-1,$$

and so

$$/2.4/ \qquad P(\rho=r)=\frac{(N-1)!}{N^{N-1}}\sum_{\nu_1+\ldots+\nu_{N-1}=r-1}\ \prod_{j=1}^{N-1}\left(\frac{j}{N}\right)^{\nu_j},$$

where $\nu_j$ are nonnegative integers.

Consequently

$$/2.5/ \qquad P(\rho-1=r-1)=\sum_{\nu_1+\ldots+\nu_{N-1}=r-1}\ \prod_{j=1}^{N-1}\left(\frac{j}{N}\right)^{\nu_j}\left(1-\frac{j}{N}\right).$$

Let now $\eta=\sum_{j=1}^{N-1}\theta\left(\frac{j}{N}\right)$. Then

$$/2.6/ \qquad P(\eta=K)=\sum_{\nu_1+\ldots+\nu_{N-1}=K}\ \prod_{j=1}^{N-1}\left(\frac{j}{N}\right)^{\nu_j}\left(1-\frac{j}{N}\right),$$

i.e. $P(\eta=K)=P(\rho-1=K)$, and so /1.10/ holds.

Now we prove /1.12/. Since the characteristic function of $e^{-e^{-x+1}}$ is $e^{-i\kappa}\Gamma(1-i\kappa)$, therefore it is enough to prove that $\psi_{\eta_N}(\kappa)\to\Gamma(1-i\kappa)e^{-i\kappa}$ uniformly in every bounded interval. Here $\psi_\xi(\kappa)$ denotes the characteristic function of $\xi$. For the sake of brevity let $\psi_j(\tau)=\psi_{\theta(\frac{j}{N})}(\tau)$.

Since

$$\Psi_j(\tau) = \sum_{\ell=0} \left(\frac{N-j}{N}\right) \left(\frac{j}{N}\right)^{\ell} e^{i\ell\tau} = \frac{N-j}{N-j\cdot e^{i\tau}} ,$$

we have

$$\Psi_\eta(\tau) = \prod_{j=1}^{N-1} \frac{N-j}{N-j\cdot e^{i\tau}} .$$

It is clear that for $\xi_N = \frac{\eta - N\log N}{N}$, we have

$$\Psi_{\xi_N}(\kappa) = e^{-i\kappa\log N} \prod_{j=1}^{N-1} \frac{N-j}{N-j \bullet \exp(i\kappa/N)} .$$

By the notation $w = 1 - \exp(i\kappa/N)$ we have

$$\frac{N-j}{N-j\cdot e^{i\frac{\kappa}{N}}} = \frac{N-j}{N-j-jw} = \frac{1}{1+\frac{j}{N-j}w} = \frac{1}{1-w}\cdot\frac{1}{1+\frac{Nw}{(1-w)(N-j)}} .$$

Assuming that $\kappa$ is bounded,

$$\Psi_{\xi_N}(\kappa) = (1-w)^{-(N-1)} e^{-i\kappa\log N} \prod_{j=1}^{N-1} \frac{1}{1+\frac{Nw}{(N-j)(1-w)}} .$$

By using the well known relations

$$\frac{1}{\Gamma(z)} = e^{\gamma z}\cdot z \prod_{\nu=1}^{\infty} \left(1+\frac{z}{\nu}\right) \exp\left(-\frac{z}{\nu}\right), \quad \Gamma(z+1) = z\cdot\Gamma(z)$$

for the Euler gamma function, we have

$$\Psi_{\xi_N}(\kappa) = (1+o(1))\,(1-w)^{N-1}\; e^{-i\kappa \log N}\; e^{\gamma\frac{Nw}{1-w}} \cdot e^{-\frac{Nw}{1-w}\sigma_N}\,\Gamma\Big(1 + \frac{Nw}{1-w}\Big),$$

where $\gamma$ being Euler's constant, $\sigma_N = \sum_{j=1}^{N-1} \frac{1}{j}$.

Since

$$\frac{Nw}{1-w} = Nw + O\Big(\frac{1}{N}\Big) = -i\kappa + O\Big(\frac{1}{N}\Big),$$

$$(1-w)^{-(N-1)} = e^{Nw} + O\Big(\frac{1}{N}\Big) = e^{-i\kappa} + O\Big(\frac{1}{N}\Big),$$

$$\sigma_N = \log N + \gamma + o(1),$$

therefore

$$\Psi_{\xi_N}(\kappa) = (1 + o(1))\,\Gamma(1 - i\kappa)\, e^{-i\kappa}.$$

Consequently

$$\lim_{N\to\infty} P(\xi_N < x) = e^{-e^{-x+1}}$$

Since $\rho = \eta + 1$, therefore $0 < \eta_N - \xi_N = \frac{1}{N}$, and so /1.12/ holds too. Theorem 1 has been proved.

§3. Proof of Theorem 2. Let $M \in [1,N]$. It is obvious that the events $\{\Lambda(M) > r\}$ and $\{s_r < M\}$ are equivalent. So

$$P(\Lambda(M) > r) = P(s_r < M) = \sum_{h=1}^{M-1} P(s_r = h).$$

Let

$$X_h = \sum_{j=1}^{h} \theta\left(\frac{j}{N}\right),$$

$\theta(\cdot)$ was defined in the first section. We have

$$P(s_r=h) = \sum_{\substack{k_1,\ldots,k_r \\ s_r=h}} p(k_1,\ldots,k_r) =$$

$$= \frac{h}{N}\,\gamma(h-1,N) \sum_{s_1\leq \ldots \leq s_{r-1}\leq h} \frac{s_1}{N}\cdots\frac{s_{r-1}}{N} =$$

$$= \frac{h}{N\left(1-\frac{h}{N}\right)}\,\gamma(h,N) \sum_{s_1\leq \ldots\leq s_{r-1}\leq h} \frac{s_1}{N}\cdots\frac{s_{r-1}}{N} =$$

$$= \frac{h}{N-h} \sum_{\nu_1+\ldots+\nu_h=r-1} \prod_{j=1}^{h} \left(\frac{j}{N}\right)^{\nu_j}\left(1-\frac{j}{N}\right) = \frac{h}{N-h}\, P(X_h=r-1),$$

if $h<N$.

So we have

$$P(\Lambda(M) > r) = \sum_{h=1}^{M-1} \frac{h}{N-h}\, P(X_h = r-1). \qquad /3.1/$$

It is easy to see that

$$M\,\theta(p) = \frac{p}{q}, \quad D^2\,\theta(p) = \left(\frac{p}{q}\right)^2 + \frac{p}{q}, \qquad /3.2/$$

where

$$D^2\xi = M\xi^2 - (M\xi)^2,$$

the variance of $\xi$ , furthermore that

/3.3/ $$M\left|\theta(p)-\frac{p}{q}\right|^3=O\left(\left(\frac{p}{q}\right)^3\right).$$

Let $A_\ell = M X_\ell = \sum_{j=1}^{\ell} \frac{j}{N-j}$ , $D_\ell^2 = D^2 X_\ell = \sum_{j=1}^{\ell} \left(\frac{j}{N-j}\right)^2 +$
$+ \sum_{j=1}^{\ell} \frac{j}{N-j}$ .

Let $\alpha \in (0,1)$ be fixed, $M = [\alpha N]$, $N \to \infty$. Let A be an arbitrary large but fixed number, $|z| < A$. Assume that $\varepsilon_N \to 0$, $\varepsilon_N \geq M^{-1/4}$ . We put $r_o = [A_M + zD_M]$, $R_1 = r_o - [\varepsilon_N D_M] +$ $+ 1$, $R_2 = r_o + [\varepsilon_N D_M] + 1$.

Let

$$A(M;\ R_1,R_2) = \sum_{r=R_1+1}^{R_2+1} P(\Lambda(M) > r).$$

From /3.1/ we have

/3.4/ $$A(M;R_1,R_2) = \sum_{h=1}^{M-1} \frac{h}{N-h} P(X_h \in [R_1,R_2]).$$

Furthermore

/3.5/ $$P(\Lambda(M) > R_1) \geq \frac{A(M;R_1,R_2)}{R_2-R_1+1} \geq P(\Lambda(M) > R_2).$$

Now we give the asymptotic of /3.4/. First of all we prove that the contribution of the term $P(X_h \in [R_1,R_2])$ is small if $h < M-M^{1/2}\varepsilon_N^{-2}$. Indeed, by using the Tshebysheff inequality,

$$P\left(X_h \in [R_1,R_2]\right) \leq P\left(X_h - A_h \geq R_1 - A_h\right) \leq \frac{D_t^2}{(R_1-A_h)^2}.$$

Since $D_h^2 = O(h)$, $A_M - A_h = \sum_{j=h+1}^{M} \frac{j}{N-j}$, therefore

$R_1 - M_h = A_M - A_h + (z - \varepsilon_N) D_M \gg M-h$, if $\frac{M-h}{M^{1/2}} \to \infty$ . So for a sufficient large $N$ and $U = M^{1/2}\varepsilon_N^{-2}$ we have

$$\frac{1}{R_2 - R_1 + 1} \sum_{h \leq M-U} \frac{h}{N-h} P\left(X_h \in (R_1,R_2)\right) \leq$$

/3.6/

$$\leq \frac{O(1)}{2\,\varepsilon_N D_M} \sum_{h \leq M-U} \frac{h}{(M-h)^2} \ll \frac{1}{\varepsilon_N M^{1/2}} \frac{M}{U} = O(\varepsilon_N) \to 0 .$$

Let now $h \in (M-U, M)$.

We shall use the theorem of Ljapunov /see [2]/, that we quote as

LEMMA 1. Let $Y_1, \ldots, Y_\ell$ be independent random variables, let

$$MY_i = 0, \quad MY_i^2 = \sigma_i^2, \quad M|Y_i|^3 = \beta_i \quad (i=1,\ldots),$$

$$S_\ell^2 = \sum_{i=1}^{\ell} \sigma_i^2, \quad B_2 = \frac{1}{\ell} S_\ell^2; \quad B_3 = \frac{1}{\ell} \sum_{i=1}^{\ell} \beta_i,$$

$$\bar{F}(x) = P\left(Y_1 + \ldots + Y_\ell < x\, S_\ell\right).$$

Then

$$\left|\tilde{F}(x) - \Phi(x)\right| < \frac{c}{\sqrt{\ell}} \quad \frac{B_3}{B_2^{3/2}} ,$$

where c is an absolute constant.

By using the relations /3.2/, /3.3/ for $Y_i = \theta\left(\frac{i}{N}\right) - M\,\theta\left(\frac{i}{N}\right)$, we have

$$\left|P\left(X_\ell - A_\ell < xD_\ell\right) - \Phi(x)\right| < \frac{c_1}{\sqrt{\ell}} \le \frac{c_2}{\sqrt{M}} ,$$

if $\ell \in (M-U, M)$, $c_1$, $c_2$ being positive constants.

So we have that

$$P\left(X_\ell \in [R_1, R_2]\right) = \Phi\left(\frac{R_2 - A_\ell}{D_\ell}\right) - \Phi\left(\frac{R_1 - A_\ell}{D_\ell}\right) + O\left(\frac{1}{\sqrt{M}}\right).$$

Consequently hence and from /3.6/ we have

$$\frac{A(M; R_1, R_2)}{R_2 - R_1 + 1} = \frac{1}{R_2 - R_1 + 1} \sum_{\ell = M-U} \frac{\ell}{N-\ell}\left(\Phi\left(\frac{R_2 - A_\ell}{D_\ell}\right) - \Phi\left(\frac{R_1 - A_\ell}{D_\ell}\right)\right) +$$

$$+ O(\varepsilon_N) + O\left(\frac{1}{\varepsilon_N M}\right).$$

Now we estimate the sum

$$B \overset{\text{def}}{=} \sum_{\ell = M-U}^{M} \frac{\ell}{N-\ell}\left(\Phi\left(\frac{R_2 - A_\ell}{D_\ell}\right) - \Phi\left(\frac{R_1 - A_\ell}{D_\ell}\right)\right). \tag{3.7}$$

We have

$$\frac{\ell}{N-\ell} = \frac{\alpha}{1-\alpha} + O\left(\frac{M-\ell}{M}\right), \quad \left(\frac{\ell}{N-\ell}\right)^2 = \left(\frac{\alpha}{1-\alpha}\right)^2 + O\left(\frac{M-\ell}{M}\right),$$

and so

$$A_M - A_\ell = \frac{\alpha}{1-\alpha}(M-\ell) + 0\left(\frac{(M-\ell)^2}{M}\right),$$

$$D_M^2 - D_\ell^2 = \left[\left(\frac{\alpha}{1-\alpha}\right)^2 + \left(\frac{\alpha}{1-\alpha}\right)\right](M-\ell) + 0\left(\frac{(M-\ell)^2}{M}\right).$$

Furthermore

$$0 \le \frac{1}{D_\ell} - \frac{1}{D_M} = 0\left(\frac{M-\ell}{M^{3/2}}\right).$$

So we have

$$\frac{R_2 - A_\ell}{D_\ell} = \left[(A_M - A_\ell) + (z+\varepsilon_N)D_M + 0(1)\right]\left[\frac{1}{D_M} + 0\left(\frac{M-\ell}{M^{3/2}}\right)\right] =$$

$$= \frac{\alpha}{1-\alpha} \quad \frac{M-\ell}{D_M} + (z+\varepsilon_N) + 0\left(\frac{(M-\ell)^2}{M^{3/2}} + \frac{M-\ell}{M}\right),$$

and similarly that

$$\frac{R_1 - A_\ell}{D_\ell} = \frac{\alpha}{1-\alpha} \cdot \frac{M-\ell}{D_M} + (z-\varepsilon_N) + 0\left(\frac{(M-\ell)^2}{M^{2/3}} + \frac{M-\ell}{M}\right).$$

Since $\phi(u+y) - \phi(u) = 0(y)$, and by /3.7/ we have

$$B = \frac{\alpha}{1-\alpha} \sum_{\ell=M-U}^{M} \phi\left(\frac{\alpha}{1-\alpha} \quad \frac{M-\ell}{D_M} + (z+\varepsilon_N)\right) -$$

$$- \phi\left(\frac{\alpha}{1-\alpha} \quad \frac{M-\ell}{D_M} + (z-\varepsilon_N)\right) + \Delta,$$

where

$$\Delta \ll \sum_{\ell=M-U}^{M} \frac{(M-\ell)^2}{M^{3/2}} + \frac{M-\ell}{M} \ll \frac{U^3}{M^{3/2}} + \frac{U^2}{M} \ll \frac{1}{\varepsilon_N^{\,6}} .$$

By using the monotonity of $\phi(\cdot)$ , we have

$$\sum_{\ell=M-U}^{M} \phi\left(\frac{\alpha}{1-\alpha} \frac{M-\ell}{D_M} + (z+\varepsilon_N)\right) = \int_0^U \phi\left(\frac{\alpha}{1-\alpha} \frac{x}{D_M} + z + \varepsilon_N\right) dx =$$

$$= \frac{1-\alpha}{\alpha} D_M \int_0^{\tau} \phi\left(Y+z+\varepsilon_N\right) dY + O(1),$$

where $\tau = \frac{\alpha}{1-\alpha} \frac{U}{D_M}$ ; and similarly that

$$\text{/3.8/} \quad \sum_{\ell=M-U}^{M} \phi\left(\frac{\alpha}{1-\alpha} \frac{M-\ell}{D_M} + z - \varepsilon_N\right) = \frac{1-\alpha}{\alpha} D_M \int_0^{\tau} \phi\left(Y+z-\varepsilon_N\right) dY + O(1).$$

Hence we get

$$B = D_M \int_{z+\tau-\varepsilon_N}^{z+\tau+\varepsilon_N} \phi(Y)\, dY - \int_{z-\varepsilon_N}^{z+\varepsilon_N} \phi(Y)\, dY + O(\varepsilon_N^{-6}).$$

Since $\tau \to \infty$ $(N \to \infty)$, therefore

$$\frac{1}{2\varepsilon_N} \int_{z+\tau-\varepsilon_N}^{z+\tau+\varepsilon_N} \phi(Y)\, dY \to 1,$$

furthermore

$$\frac{1}{2\varepsilon_N} \int_{z-\varepsilon_N}^{z+\varepsilon_N} \phi(Y)\, dY \to \phi(z).$$

So, from /3.8/ $\frac{B}{R_2-R_1+1} = \frac{B}{2\varepsilon_N D_M + O(1)} \to 1 - \phi(z)$ , if $\frac{1}{\varepsilon_N^{\,6} M^{1/2}} \to 0$ $(N \to \infty)$, i.e. if $\varepsilon_N \to 0$ sufficiently slowly.

So we have proved that

$$\frac{A(M, R_1, R_2)}{R_2 - R_1 + 1} \to 1 - \Phi(z) \quad (N \to \infty.$$

From /3.5/ we get immediately that

$$\lim_{M \to \infty} P\left(\frac{\Lambda(M) - A_M}{D_M} \geq z - \delta\right) \geq 1 - \Phi(z),$$

$$\lim_{M} P\left(\frac{\Lambda(M) - A_M}{D_M} \geq z + \delta\right) \leq 1 - \Phi(z)$$

for every $z$ and every fixed $\delta > 0$. This gives that

$$\lim_{M \to \infty} P\left(\frac{\Lambda(M) - A_M}{D_M} \geq z\right) = 1 - \Phi(z)$$

for every z.

Now we observe that

$$\frac{A_M}{N} = -\alpha + \sum_{j=1}^{M} \frac{1}{N-j} = -\alpha + \left(\log N - \log(N-M)\right) + O\left(\frac{1}{N}\right) =$$

$$= -\alpha - \log(1-\alpha) + O\left(\frac{1}{N}\right),$$

$$\frac{D_M^2}{N} = \sum_{j=1}^{M} \left(\frac{j}{N-j}\right)^2 + \frac{A_M}{N} = \log(1-\alpha) + \left(\frac{1}{1-\alpha} - 1\right) + O\left(\frac{1}{N}\right).$$

Hence our theorem follows immediately.

§4. Proof of Theorem 4. Let $\varphi_h(z) := \sum_{\ell=0}^{\infty} P(X_h=\ell)z^{\ell}$.

Since $X_h = \sum_{j=1}^{h} \theta\left(\frac{j}{N}\right)$, we have

$$\varphi_h(z) = \prod_{j=1}^{h} \frac{N-j}{N-jz} .$$

So for $|z| \leq 1$, $h = 0(\sqrt{N})$ we have

$$\varphi_h(z) = \prod_{j=1}^{h} \frac{1}{\frac{j}{N-j}(1-z)} = \exp\left( -A_h (1-z)\right) + 0\left((1-z)^2 \sum_{j=1}^{h} \left(\frac{j}{N-j}\right)^2\right) =$$

$$= \exp\left(A_h (z-1)\right) + 0\left(\frac{h^3}{N^2}\right).$$

We take $\Phi_h(z) = \exp\left(A_h (z-1)\right)$. By the Parseval formula we deduce that

$$P\left(X_h = \ell\right) = \int_{-1/2}^{1/2} \varphi_h\left(e^{2\pi i\alpha}\right) e^{-2\pi i\ell\alpha}\, d\alpha =$$

$$\int_{-1/2}^{1/2} \Phi_h\left(e^{2\pi i\alpha}\right) e^{-2\pi i\ell\alpha}\, d\alpha + 0\left(\frac{h^3}{N^2}\right).$$

Since the integral on the right hand is $\frac{A_h^{\ell}}{\ell!} \exp(-A_h)$, so we have

$$P\left(X_h = \ell\right) = \frac{A_h}{\ell!} \exp\left(-A_h\right) + 0\left(\frac{h^3}{N^2}\right) .$$

We shall use relation /3.1/. Observing that

$$A_h = \frac{h^2}{2N} + 0\left(\frac{h^3}{N^2}\right) = \frac{h^2}{2N}\left(1+0\left(\frac{1}{\sqrt{N}}\right)\right), \quad \frac{h}{N-h} = \frac{h}{N} + 0\left(\left(\frac{h}{N}\right)^2\right),$$

we have

$$P(\Lambda(M) > r) = \sum_{h=1}^{M-1} \frac{h}{N-h} P(X_h = r-1) =$$

$$= \frac{1}{N} \sum_{h=1}^{M-1} h \left\{ \frac{A_h^{r-1}}{(r-1)!} \exp(-A_h) + O\left(\frac{h^3}{N^2}\right) \right\} + O\left(\frac{M^3}{N^2}\right) =$$

$$= \frac{1}{N(r-1)!} \sum_{h=1}^{M-1} h A_h^{r-1} \exp(-A_h) + O(N^{-1/2}) =$$

$$= \frac{1}{N(r-1)!} \left\{ \sum_{h=1}^{M-1} h \left(\frac{h^2}{2N}\right)^{r-1} \exp\left(-\frac{h^2}{2N}\right) \right\} \left(1 + O\left(\frac{r}{\sqrt{N}}\right)\right) + O(N^{-1/2}).$$

Furthermore

$$\sum_{h=1}^{M-1} h \left(\frac{h^2}{2N}\right)^{r-1} \exp\left(-\frac{h^2}{2N}\right) = \left(1 + O\left(\frac{1}{\sqrt{N}}\right)\right) \int_0^M \left(\frac{x^2}{2N}\right)^{r-1} x \exp\left(-\frac{x^2}{2N}\right) dx =$$

$$= \left(1 + O\left(\frac{1}{\sqrt{N}}\right)\right) N \int_0^{\frac{M^2}{2N}} Y^{r-1} \cdot \exp(-Y)\, dY.$$

Since $\frac{M^2}{2N} \to \frac{A^2}{2N}$, we have

$$\lim_{N\to\infty} P(\Lambda(M) > r) = \frac{1}{(r-1)!} \int_0^{A^2/2} Y^{r-1} \exp(-Y)\, dY,$$

if $r \geq 1$.

Observing that $P(\Lambda(M) > 0) = 1$, and

$$P(\Lambda(M) = r) = P(\Lambda(M) > r-1) - P(\Lambda(M) > r),$$

we have

$$\lim_{N\to\infty} P\left(\Lambda(M) = r\right) = \int_0^{A^2/2} \left\{ \frac{1}{(r-2)!} Y^{r-2} - \right.$$

$$\left. - \frac{1}{(r-1)!} Y^{r-1} \right\} \exp(-Y)\, dY = \exp\left(- \frac{A^2}{2}\right)\cdot\left(\frac{A^2}{2}\right)^{r-1} \frac{1}{(r-1)!} ,$$

if $r \geq 2$. To prove the assertion for $r=1$ we observe that

$$P\left(\Lambda(M) = 1\right) = 1 - P\left(\Lambda(M) > 1\right) =$$

$$= 1 - \int_0^{A^2/2} \exp(-Y)\, dY = 1 - \exp\left(- \frac{A^2}{2}\right).$$

## References


[1] A. Iványi and I. Kátai, Processing of random sequences with priority, Acta Cybernetica, 4 /1975/, 85-101.

[2] M. Loéve, Probability theory, New York, 1978. Springer-Verlag.

# ON THE EXISTENCE OF MINIMAL COMPLETE CLASSES OF ESTIMATORS

Andrzej Kozek, Wrocław
Wolfgang Wertz, Wien

## 0. Summary and introduction.

In [9] Pitcher introduced a notion of compactness of statistical structures. He showed, among others, that the compactness implies the existence of a minimal complete class of estimators. Pitcher assumed that the loss function $L(x,P)$ is of the form

$$L(x,P) = |x - g(P)|^p , \qquad 1 < p < \infty ,$$

where $x \in R^1$, $g(\cdot)$ is a bounded function on a considered set of probability measures. Moreover, he assumed that the set of considered estimators consists of those functions which have norms in $L_p(P)$ uniformly bounded with respect to P. Pitcher showed that $\sigma$-dominated statistical structures are compact whereas Kusama and Yamada [7] showed that discrete statistical structures are compact.

The aim of this paper is to show that the compactness implies the existence of minimal complete classes of estimators for a large class of convex loss functions and for more natural classes of estimators taking values in a separable dual Banach space which need not have uniformly bounded norms. This result improves, e.g., Theorem 1 in Wertz [13], where norms in Orlicz spaces instead of risk functions were considered as a measure of optimality of estimators of density functions. It is also shown that statistical structures dominated by a decomposable measure are compact. This implies both the compactness in the $\sigma$-dominated and in the discrete cases.

Moreover, let us point out that the existence of minimal complete classes consisting of non-randomized decision functions can be also derived starting from LeCam's theory of decision functions [8]. A schema of reasoning is then the following one:

1. take a Čech-Stone compactification $\hat{X}$ of the decision space $X$, 2. infer from LeCam's theory the existence of a minimal complete class of behaviouristic decision rules concentrated on $\hat{X}$, 3. find conditions for the validity of the statement on completeness of the class of decision rules concentrated on $X$, 4. infer that if the loss function is convex, then the nonrandomized decision rules form a complete class, 5. conclude that there exists a minimal complete class of estimators consisting of non-randomized estimators. (For steps 3 and 4 see [2]).

In our case the decision space $X$ is convex (it is a Banach space, even) and the risk function is a convex and lower semicontinuous functional on a vector space of estimators. Thus, there is no reason to consider any randomization which aims in providing convexity of the set of decision rules and convexity and lower semicontinuity of the risk function. Clearly, the topologies considered in our case are different from those considered by LeCam in [8] and are related to the topologies considered in the theory of Orlicz spaces.

1. Preliminaries and notation.

1.1. Let $(T,\mathcal{A})$ be a measurable space and $P$ a probability measure on $\mathcal{A}$. Let $\mathcal{A}_P = \{A \subset T$: there exists $A_1 \in \mathcal{A}$ such that $P^*(A_1 \Delta A) = 0\}$, where $P^*$ stands for the outer measure generated by $P$. $\mathcal{A}$ is called $P$-complete whenever $\mathcal{A} = \mathcal{A}_P$. Let $\mathcal{P}$ be a class of probability measures on $(T,\mathcal{A})$ and put $\mathcal{A}_{\mathcal{P}} = \{A \subset T$: for each $P \in \mathcal{P}$ there exists $A_P \in \mathcal{A}$ such that $P^*(A_P \Delta A) = 0\}$. $\mathcal{A}$ is called $\mathcal{P}$-complete if $\mathcal{A} = \mathcal{A}_{\mathcal{P}}$ (compare [9; p.601]). Throughout the paper we assume that the considered $\sigma$-fields are $P$-complete when we deal with a single probability measure $P$ and $\mathcal{P}$-complete when a class $\mathcal{P}$ of probability measures is considered.

1.2. Let $X$ be a Banach space and let $M(X,\mathcal{A})$ be the set of all strongly $\mathcal{A}$-measurable functions from $T$ into $X$. We shall identify functions from $M(X,\mathcal{A})$ such that $P(x_1(t) \neq x_2(t)) = 0$ for each $P \in \mathcal{P}$.

1.3. Let $\Phi$ be an N-function on $X$ (see [3,4]), where $X$ is a separable dual of a Banach space $Y$, i.e. $Y' = X$. The Young function complementary to $\Phi$ (called also the conjugate function of $\Phi$) is defined on $Y$ and given by

$$\Psi(y) = \sup_{x \in X} ((x,y) - \Phi(x)),$$

where $(x,y) = x(y)$. It is known that if $\Phi$ is an N-function, then $\Psi$ is an N-function, too [3; Proposition

4.6]. Put

$$I_{\Phi,P}(x(\cdot)) = \int \Phi(x(t))P(dt),$$

$$B_{\Phi,P}(X) = \{x(\cdot) \in M(X,\mathcal{A}_P): I_{\Phi,P}(x(\cdot)) \leq 1\}$$

and

$$L_{\Phi,P}(X) = \text{Lin } B_{\Phi,P}(X),$$

where Lin(·) stands for the linear space spanned over the indicated subset of a vector space. $L_{\Phi,P}(X)$ is called an Orlicz space and $B_{\Phi,P}(X)$, which is a convex set, stands for a unit ball in $L_{\Phi,P}(X)$. It is known that $L_{\Phi,P}(X)$ is a Banach space [3; Theorem 2.4] whenever functions equal P-a.e. to each other are identified. Denote by $N_{\Phi,P}$ the norm in $L_{\Phi,P}(X)$. We shall write $\mathcal{L}_{\Phi,P}(X)$ when individual functions instead of the equivalence classes are considered. Spaces $L_{\Psi,P}(Y)$ and $\mathcal{L}_{\Psi,P}(Y)$ are defined in a similar manner.

<u>1.4.</u> Let

$$\mathcal{L}_{\Phi,\mathcal{P}}(X) = \bigcap_{P \in \mathcal{P}} \mathcal{L}_{\Phi,P}(X)$$

and denote by $L_{\Phi,\mathcal{P}}(X)$ the vector space of the equivalence classes of functions described in <u>1.2.</u> We shall consider $L_{\Phi,\mathcal{P}}(X)$ endowed with the locally convex topology determined on $L_{\Phi,\mathcal{P}}(X)$ by seminorms $N_{\Phi,P}$.

<u>1.5.</u> From now <u>we assume $\Psi$ to be bounded above on balls of radius $n$</u>, i.e. $\Psi(y) \leq C_n$ whenever $\|y\| \leq n$. Then $E_{\Psi,P}(Y) = \{ y(\cdot) \in L_{\Psi,P}(Y): I_{\Psi,P}(k\, y(\cdot)) < \infty$ for every $k > 0\}$ is a non-trivial Banach subspace of $L_{\Psi,P}(Y)$. In fact, $E_{\Psi,P}(Y)$ is the closure (for the norm topology in $L_{\Psi,P}(Y)$) of the space of step functions, or equivalently, of the space of norm bounded functions from $M(Y,\mathcal{A}_P)$,

cf. [3;Propositions 3.2 and 3.3]. Moreover, $L_{\Phi,P}(X)$ is isometrically isomorphic to $(E_{\Psi,P}(Y))'$, i.e. for each continuous linear functional $u$ on $E_{\Psi,P}(Y)$ there is a unique $x_u(\cdot) \in L_{\Phi,P}(X)$ such that $u(y(\cdot))= = \int (x_u(t),y(t))P(dt)$ holds for every $y(\cdot) \in E_{\Psi,P}(Y)$. Note that since $I_{\Psi,P}$ is continuous on $E_{\Psi,P}(Y)$ ($I_{\Psi,P}$ is finite, convex and l.s.c. on $E_{\Psi,P}(Y)$ [3; Theorem A5,4; Theorem 1.2], hence it is continuous [10; Corollary 7C]) the convex functional $I_{\Phi,P}$ is inf-compact on $L_{\Phi,P}(X)$ endowed with the weak-star topology, i.e. for each $\alpha \in R$ set $\{x(\cdot): I_{\Phi,P}(x(\cdot)) \leq \alpha\}$ is compact for the $\sigma(L_{\Phi,P}(X), E_{\Psi,P}(Y))$-topology.

<u>1.6.</u> Note that if $P \in \mathcal{P}$ and $y(\cdot) \in E_{\Psi,P}(Y)$, then formula

$$l(y(\cdot),P)(x(\cdot)) = \int (x(t),y(t))P(dt)$$

determines a linear continuous functional on $L_{\Phi,\mathcal{P}}(X)$. Indeed, we have

$$|l(y(\cdot),P)(x(\cdot))| \leqslant N_{\Phi,P}(x(\cdot))N^1_{\Psi,P}(y(\cdot)),$$

where $N^1_{\Psi,P}$ is an Orlicz norm on $E_{\Psi,P}(Y)$. Put $E_{\Psi,\mathcal{P}}(Y) = \mathrm{Lin}\{l(y(\cdot),P): y(\cdot) \in E_{\Psi,P}(Y),\ P \in \mathcal{P}\}$.

Since both $L_{\Phi,\mathcal{P}}(X)$ and $E_{\Psi,\mathcal{P}}(Y)$ contain step functions and $X$ and $Y$ are separable the pair $(L_{\Phi,\mathcal{P}}(X), E_{\Psi,\mathcal{P}}(Y))$ endowed with the pairing

$$(x(\cdot),\ l(y(\cdot),P)) = \int (x(t),y(t))P(dt)$$

forms a dual pair and the $\sigma(L_{\Phi,\mathcal{P}}(X), E_{\Psi,\mathcal{P}}(Y))$-topology is Hausdorff.

## 2. Compact statistical structures.

<u>2.1.</u> Let us consider sets $I_g^{\leq K} = I^{\leq K}_{\Phi,\mathcal{P},g} =$

$= \{ x(\cdot) \in L_{\Phi,\mathcal{P}}(X): I_{\Phi,P}(x(\cdot) - g(P)) \leqslant K(P)$ for each $P \in \mathcal{P}\}$, where $g: \mathcal{P} \longrightarrow X$ and $K: \mathcal{P} \longrightarrow R_+$ are given mappings. We are interested in finding sufficient conditions for $I_g^{\leqslant K}$ or $I^{\leqslant K}$ $( = I_0^{\leqslant K})$ to be compact for $\sigma(L_{\Phi,\mathcal{P}}(X), E_{\Psi,\mathcal{P}}(Y))$-topology. Let us note that by a Rockafellar's Lemma (see [3; Theorem A5] or [4; Theorem 1.2]) convex functionals $I_{\Phi,P}$ considered on $L_{\Phi,\mathcal{P}}(X)$ are lower semicontinuous for the $\sigma(L_{\Phi,\mathcal{P}}(X), E_{\Psi,\mathcal{P}}(Y))$-topology. For every $P \in \mathcal{P}$ we denote by $\mathrm{lev}_K I_{\Phi,P,g}$ the level set of $I_{\Phi,P,g}$, where

$$I_{\Phi,P,g}(x(\cdot)) = \int \Phi(x(t)-g(P))P(dt),$$

$$x(\cdot) \in L_{\Phi,P}(X),$$

i.e.

$$\mathrm{lev}_K I_{\Phi,P,g} = \{ x(\cdot) \in L_{\Phi,P}(X): I_{\Phi,P,g}(x(\cdot)) \leqslant K(P) \text{ for each } P \in \mathcal{P}\} .$$

Since $I_{\Psi,P}$ is continuous on $E_{\Psi,P}(Y)$ a theorem of Moreau and Rockafellar (cf. [10; Theorem 7A(b)]) implies that the sets $\mathrm{lev}_K I_{\Phi,P,g}$ are $\sigma(L_{\Phi,P}(X), E_{\Psi,P}(Y))$-compact for each $P \in \mathcal{P}$ .

Let

$$\mathrm{prod}_{K,g} = \mathrm{prod}_{\Phi,\mathcal{P},K,g} = \prod_{P \in \mathcal{P}} \mathrm{lev}_K I_{\Phi,P,g} .$$

The space $\mathrm{prod}_{K,g}$ endowed with the Tichonov product topology is a compact space provided each set $\mathrm{lev}_K I_{\Phi,P,g}$ is endowed with the $\sigma(L_{\Phi,P}(X), E_{\Psi,P}(Y))$-topology. Let $\mathrm{diag}\, x(\cdot)$ stands for the value of the diagonal mapping from $I_g^{\leqslant K}$ into $\mathrm{prod}_{K,g}$ , i.e.

$$\mathrm{diag}\, x(\cdot) = (x(\cdot)_P)_{P \in \mathcal{P}}$$

where $x(\cdot)_P$ stands for the equivalence class in $L_{\Phi,P}(X)$ of any function from the equivalence class of $x(\cdot) \in L_{\Phi,\mathcal{P}}(X)$.

2.2. Proposition. $I_g^{\leq K}$ is compact for the $\sigma(L_{\Phi,\mathcal{P}}(X), E_{\Psi,\mathcal{P}}(Y))$-topology if and only if diag $(I_g^{\leq K})$ is closed in the product topology of $\mathrm{prod}_{K,g}$.

The proof of this Proposition is standard, see [9; p.599] and [11; Lemma 2.6].

2.3. Proposition. If

1. $(x(\cdot)_P)$ is in the closure of diag $(I^{\leq K})$, then

2. for every finite set $P_1, P_2, \ldots, P_k$ from $\mathcal{P}$ there exists $x(\cdot) \in M(X,\mathcal{A})$ such that $x(t) = x(t)_{P_i}$ $P_i$ - a.e. for $i = 1,\ldots,k$,

and

3. for every countable set $P_1, P_2, \ldots$ from $\mathcal{P}$ there exists $x(\cdot) \in M(X,\mathcal{A})$ such that $x(t) = x(t)_{P_i}$ $P_i$ - a.e. for $i = 1,2,\ldots$ .

Moreover, if for each $P \in \mathcal{P}$ there is a set $A_P \in \mathcal{A}$ such that $P(A_P) = 0$ and for any $P' \in \mathcal{P}$ its singular part with respect to $P$, say $P'_S$, is concentrated on $A_P$, i.e. $P'_S(A_P) = P'_S(T)$, then the function $x(\cdot)$ can be choosen from $I_g^{\leq K}$ and statements 1, 2 and 3 imply each other.

Proof. Clearly, it is enough to show that the first assertion implies the third one. First, we note that a similar argument as in [9; Lemma 1.2] shows that if $P'$, $P'' \in \mathcal{P}$ , then $x(t)_{P'} = x(t)_{P''}$ $P'$-a.e. and $P''$-a.e. on set $\{t \in T: dP_1/d(P_1+P_2)(t) > 0,\ dP_2/d(P_1+P_2)(t) > 0\}$. Now, let sequence $P_1, P_2, \ldots$ be given, where $P_i \in \mathcal{P}$. Let $m = \sum_{i=1}^{\infty} 2^{-i}P_i$ and let $g_i(\cdot)$, $i = 1,2,\ldots$ , stand for the indicator functions of sets

$\{t: dP_1/dm(t) > 0\}$ if $i = 1$ and

$\{t: dP_i/dm(t) > 0,\ g_1(t) = 0, \ldots, g_{i-1}(t) = 0\}$ if $i > 1$.

Let $x(\cdot)_{P_i}$ stand for individual functions and put $x(t) = \sum_{i=1}^{\infty} g_i(t)x(t)_{P_i}$. For any fixed $k$ we denote

$A_{k,i} = \{t: dP_k/dm(t) > 0,\ dP_i/dm(t) > 0\}$.

Since $P_k(\bigcup_{i=1}^{\infty} A_{k,i}) = 1$ and for any $i$ we have $x(t)_{P_k} = x(t)_{P_i}$ $P_k$-a.e. on $A_{k,i}$, we conclude $x(t) = x(t)_{P_k}$ $P_k$-a.e.

For the proof of the second part of the Proposition assume that $A_i$, $i = 1,2,\ldots$ , are the sets for which $P_i(A_i) = 0$ and $P_S'(A_i) = 0$ for $P' \neq P_i$, $P' \in \mathcal{P}$. Let $g(\cdot)$ stand for the indicator function of the set $T \setminus \bigcup_{i=1}^{\infty} A_i$ and put

$$x'(t) = g(t)\cdot x(t),$$

where $x(t)$ is the function defined in the first part of the proof. We show that $x'(\cdot) \in I^{\leq K}$. Indeed, given $P \in \mathcal{P}$ we have

$$\begin{aligned} I_{\Phi,P}(x(\cdot)) &= \int \Phi(x(t))P(dt) = \\ &= \int \Phi(g(t)\sum_i g_i(t)x(t)_{P_i})P(dt) = \\ &= \sum_i \int g(t)g_i(t)\Phi(x(t)_{P_i})P(dt) = \\ &= \sum_i \int g(t)g_i(t)\Phi(x(t)_P)P(dt) = \\ &= \int \Phi(x(t)_P)P(dt) \leq K(P), \end{aligned}$$

i.e. $x(\cdot) \in I^{\leq K}$. Hence, statement 1 implies statement 3. Since implications $3 \Longrightarrow 2$ and $2 \Longrightarrow 1$ are obvious the proof of the Proposition is concluded.

<u>2.4. Proposition</u>. Let $B = \{x(\cdot) \in M(X, \mathcal{A}_{\mathcal{P}})$: $\|x(t)\| \leqslant 1$ P-a.e. for each $P \in \mathcal{P}\}$. On B, the $\sigma(B, L_{\infty,\mathcal{P}}(Y))$- and $\sigma(B, E_{\Psi,\mathcal{P}}(Y))$-topologies coincide for any N-function $\Psi$ bounded above on balls in Y centred at zero. In particular, topologies $\sigma(B, L_{\infty,\mathcal{P}}(Y))$ and $\sigma(B, L_{1,\mathcal{P}}(Y))$ coincide on B.

Proof. Since every bounded function is in $E_{\Psi,\mathcal{P}}(Y)$ for any $P \in \mathcal{P}$ and $\Psi$ (satisfying the assumed conditions) the $\sigma(B, L_{\infty,\mathcal{P}}(Y))$-topology is weaker than the $\sigma(B, E_{\Psi,\mathcal{P}}(Y))$-topology. Now, if $l(y(\cdot),P) \in E_{\Psi,\mathcal{P}}(Y)$, let

$$y_n(t) = \begin{cases} y(t) & \text{if } \|y(t)\| \leqslant n \\ 0 & \text{otherwise.} \end{cases}$$

Then $l(y_n(\cdot),P) \in L_{\infty,\mathcal{P}}(Y)$ for $n = 1,2,\dots$ and

$$|l(y(\cdot),P)(x(\cdot)) - l(y_n(\cdot),P)(x(\cdot))| = \left|\int (x(t), y(t)-y_n(t))P(dt)\right|$$

$$\leqslant \int \|y(t) - y_n(t)\| \, P(dt) = \int_{\|y_n(t)\| > n} \|y(t)\| \, P(dt),$$

which tends to 0 uniformly in $x(\cdot) \in B$. Hence, $l(y(\cdot),P)$ is on B a uniform limit of functionals from $L_{\infty,\mathcal{P}}(Y)$, so that the topologies are equivalent on B.

<u>2.5. Theorem</u>. Let $K(P) \geqslant 1$ for each $P \in \mathcal{P}$. Then $I^{\leqslant K}$ is $\sigma(L_{\Phi,\mathcal{P}}(X), E_{\Psi,\mathcal{P}}(Y))$-compact if and only if B is $\sigma(B, L_{1,\mathcal{P}}(Y))$-compact.

Proof. Let $I^{\leqslant K}$ be compact. Since $\Phi$ is an N-function there exist $r > 0$ and $s > 0$ such that $\Phi(x) \leqslant s$ whenever $\|x\| \leqslant r$. Assuming (with no loss of generality) $s \leqslant 1$ we easily get $B \subset \frac{1}{r} I^{\leqslant K}$. Since $I^{\leqslant K}$ is compact it is enough to verify that diag B is closed in $\mathrm{diag}(\frac{1}{r} I^{\leqslant K})$.

Let $(x(\cdot)_P) \in \text{diag}\,(\frac{1}{r}\, I^{\leqslant K})$ and $(x(\cdot)_P) \in \text{cl diag}\, B$. By the compactness of $\text{diag}\,(\frac{1}{r}\, I^{\leqslant K})$ there exists $x(\cdot) \in \frac{1}{r}\, I^{\leqslant K}$ such that $\text{diag}\, x(\cdot) = (x(\cdot)_P)$. Suppose that there are $\gamma > 0$, $P \in \mathcal{P}$ and $A \in \mathcal{A}_{\mathcal{P}}$ such that $\| x(t) \| > 1 + \gamma$ for $t \in A$ and $P(A) = \beta$, $\beta > 0$. Take an $\mathcal{A}_{\mathcal{P}}$-measurable selector $y(t)$ of the multifunction $t \longrightarrow \{y \in Y: (x(t),y) \geqslant 1 + \gamma/2,\ \| y \| = 1\}$, $t \in A$. Note that the existence of the desired $y(\cdot)$ follows easily from Theorem 1 in Kuratowski and Ryll-Nardzewski [6] (see also Theorem 4.2. f in an excellent review paper [12]). Extend $y(\cdot)$ onto $T$ putting $y(t) = 0$ when $t \notin A$. Then $y(\cdot) \in L_{Y,P}(Y)$ and we get

$$
\begin{aligned}
1 + \gamma/2 \leqslant \frac{1}{\beta} \int_A (x(t),y(t))P(dt) &= \\
&= \int (x(t),\tfrac{1}{\beta}y(t))P(dt) = \\
&= \int (x(t)-x_B(t),\tfrac{1}{\beta}y(t))P(dt) + \\
&+ \int (x_B(t),\tfrac{1}{\beta}y(t))P(dt) \leqslant \varepsilon + 1,
\end{aligned}
$$

where $x_B(\cdot)$ is an appropriate function from $B$. Since $\varepsilon$ is arbitrary we get $1 + \gamma/2 \leqslant 1$ that is not true. Thus $P(A) = 0$ must hold and hence $x(\cdot) \in B$. So, diag $B$ is closed in $\text{diag}\,(\frac{1}{r}\, I^{\leqslant K})$ and therefore it is compact. By 2.2 this yields the compactness of $B$.

Now, assume $B$ to be compact and let $(x(\cdot)_P)$ be in the closure of $\text{diag}\, I^{\leqslant K}$. By 2.3, for all $\{P_1, P_2, \ldots, P_n\} \subset \mathcal{P}$ there is an $x'(\cdot) \in I^{\leqslant K}$ such that $x'(t) = x(t)_{P_i}$, $P_i$-a.e., $i = 1,\ldots,n$. Put

$$x_{P_i}^{(n)}(t) = \begin{cases} x_{P_i}(t) & \text{if} \quad \|x_{P_i}(t)\| \leq n \\ 0 & \text{otherwise} \end{cases}$$

and

$$x'^{(n)}(t) = \begin{cases} x'(t) & \text{if} \quad \|x'(t)\| \leq n \\ 0 & \text{otherwise.} \end{cases}$$

Then $\frac{1}{n} x_{P_i}^{(n)}(\cdot) = \frac{1}{n} x'^{(n)}(\cdot)$ $P_i$-a.e. and since $\frac{1}{n} x'^{(n)}(\cdot)$ is in $B$ we get that $(\frac{1}{n} x^{(n)}(\cdot)_P)$ is in cl diag $B$ = diag $B$ for each $n$. Hence, by 2.2, there is an $x''_n(\cdot) \in B$ such that $\frac{1}{n} x^{(n)}(\cdot)_P = x''_n(\cdot)$ $P$-a.e. for each $P \in \mathcal{P}$. Therefore, $nx''_n(t)$ is convergent $P$-a.e. for each $P \in \mathcal{P}$. Put

$$x''(t) = \begin{cases} \lim nx''_n(t) & \text{if the limit exists} \\ 0 & \text{otherwise.} \end{cases}$$

Clearly, $x''(t) = x_P(t)$ holds $P$-a.e. for each $P \in \mathcal{P}$. So, $x''(\cdot)$ is in $I^{\leq K}$ because $x_P(\cdot)$ is in $\text{lev}_K I_{\Phi,P}$ for each $P \in \mathcal{P}$. Thus, the Theorem is proved.

<u>2.6. Proposition</u>. Suppose that $\mathcal{P}' \subset \mathcal{P}$ and for each $A \in \mathcal{A}_{\mathcal{P}'}$, there is $A' \in \mathcal{A}_{\mathcal{P}}$ such that $P(A \Delta A') = 0$ for every $P \in \mathcal{P}'$. If $I_{\mathcal{P}}^{\leq K}$ is $\sigma(L_{\Phi,\mathcal{P}}(X), E_{\Psi,\mathcal{P}}(Y))$-compact and $K(P) \geq 1$ for each $P \in \mathcal{P}$, then $I_{\mathcal{P}'}^{\leq K}$ is $\sigma(L_{\Phi,\mathcal{P}'}(X), E_{\Psi,\mathcal{P}'}(Y))$-compact.

Proof. Theorem 2.5 and our assumption imply that $B$ is $\sigma(L_{\Phi,\mathcal{P}}(X), E_{\Psi,\mathcal{P}}(Y))$-compact. Let $B' = \{x(\cdot) \in M(X, \mathcal{A}_{\mathcal{P}'}) : \|x(t)\| \leq 1$ $P$-a.e. for each $P \in \mathcal{P}'\}$. The identity map from $B$ into $B'$ is continuous, so the image of $B$ in $B'$ is $\sigma(L_{\Phi,\mathcal{P}'}(X), E_{\Psi,\mathcal{P}'}(Y))$-compact.

Moreover, by our assumption, any equivalence class in $L_{\Phi,\mathcal{P}'}(X)$ contains an $\mathcal{A}_{\mathcal{P}'}$-measurable function $x(\cdot)$ such that $\|x(t)\| \leqslant 1$ holds everywhere. Hence this map is onto $B'$, thus $B'$ is compact. Now, the part "if" of Theorem 2.5 yields the compactness of $I_{\mathcal{P}'}^{\leqslant K}$.

<u>2.7. Proposition</u>. If $K(P) \geqslant 1$ and $I_{\mathcal{P}}^{\leqslant K}$ is compact, then $I_{\overline{\mathcal{P}}}^{\leqslant K}$ is compact, where $\overline{\mathcal{P}}$ is the set of probability measures $P'$ which are dominated by some countable subset of $\mathcal{P}$. In particular, if $\mathcal{P}$ is dominated, then $I_{\mathcal{P}}^{\leqslant K}$ is compact.

The proof of this Proposition is virtually the same as in Pitcher [9; Theorem 1.2].

<u>2.8. Proposition</u>. If $K\vee 1(P) = \max(K(P),1)$ and $I^{\leqslant K\vee 1}$ is compact, then $I^{\leqslant K}$ is compact.

Proof. By the lower semicontinuity of $I_{\Phi,P}$ (see <u>2.1</u>) the set $I^{\leqslant K}$ is a closed subset of $I^{\leqslant K\vee 1}$. This implies the compactness of $I^{\leqslant K}$.

<u>2.9.</u> Recall that a measure space $(T,\mathcal{A},m)$ is decomposable if and only if there is a partition $\{T_i\}_{i\in I}$ of $T$ such that 1) $T_i \in \mathcal{A}$ and $0 < m(T_i) < \infty$ for every $i \in I$, 2) given a set $A \subset T$, $A \in \mathcal{A}$ if and only if $A \cap T_i \in \mathcal{A}$ for every $i \in I$, 3) for every $A \in \mathcal{A}$ we have $m(A) = \sum_{i\in I} m(A\cap T_i)$ (the index set $I$ is here arbitrary), see [1, p.172].

<u>Proposition</u>. If $\mathcal{P}$ is dominated by a decomposable measure, then $I^{\leqslant K}$ is compact.

Proof. Let $K(P) \geqslant 1$ for each $P \in \mathcal{P}$. We have $(T_i,\mathcal{A}_i,m_i)$, $i \in I$, a collection of $m_i$-complete measure spaces, where $0 < m_i(T_i) < \infty$. By Proposition 2.7 the sets $I_{\mathcal{P}_i}^{\leqslant K}$ are compact, where $\mathcal{P}_i$ is the class of all probability measures on $(T_i, \mathcal{A}_i)$ dominated by $m_i$. The argument given in [9; Example 1] shows that if $T = \bigcup T_i$, $\mathcal{A} = \{A \subset T: A \cap T_i \in \mathcal{A}_i$ for each $i \in I\}$ and $P_i$ are defined on $\mathcal{A}$ by $P_i(A) = P_i(A \cap T_i)$, then $I^{\leqslant K}_{\bigcup_{i \in I} \mathcal{P}_i}$ is compact. By Proposition 2.7 $I^{\leqslant K}_{\overline{\bigcup_{i \in I} P_i}}$ is compact and hence Proposition 2.6 implies that for any $\mathcal{P} \subset \overline{\bigcup_{i \in I} \mathcal{P}_i}$ the set $I^{\leqslant K}_{\mathcal{P}}$ is compact for the appropriate considered topology. Now, Proposition 2.8 implies that $I^{\leqslant K}_{\mathcal{P}}$ is compact for any $K(P) \geqslant 0$. This completes the proof.

Let us note that the statement converse to Proposition 2.8 does not need be true. If, e.g., $K(P) < 1$, then it may happen that $I^{\leqslant K}$ consists of the function identically equal to zero and, simultanousely, $I^{\leqslant 1}$ may be non-compact.

<u>2.10. Proposition</u>. If $\Phi$ is finite on $X$ and $I^{\leqslant K'}$ is compact, where $K'(P) = \frac{1}{2}(K(P) + \Phi(g(P)),$ then $I_g^{\leqslant K}$ is compact.

Proof. Note that since

$$I_{\Phi,P}(\tfrac{1}{2}x(\cdot)) = I_{\Phi,P}(\tfrac{1}{2}(x(\cdot)-g(P))+\tfrac{1}{2}g(P)) \leqslant \tfrac{1}{2}I_{\Phi,P}(x(\cdot)-g(P))+ + \tfrac{1}{2}\Phi(g(P)),$$

holds, we obtain

$$I_{\Phi,P}(x(\cdot) - g(P)) \geqslant 2I_{\Phi,P}(\tfrac{1}{2}x(\cdot)) - \Phi(g(P))$$

and

$$I_g^{\leqslant K} = \{ x(\cdot) \in L_{\Phi,\mathcal{P}}(X): I_{\Phi,P}(x(\cdot)-g(P)) \leqslant K(P) \}$$

$$\subset \{ x(\cdot) \in L_{\Phi,\mathcal{P}}(X): I_{\Phi,P}(\tfrac{x(\cdot)}{2}) \leqslant$$

$\frac{1}{2}(K(P) + \Phi(g(P)))\} = 2I^{\leq K'}$.

Since, by the lower semicontinuity of $I_{\Phi,P}$, the set $I_g^{\leq K}$ is closed we get that $I_g^{\leq K}$ is compact.

In view of sections 2.4 and 2.5 where we have shown some independence of the compactness of $I^{\leq K}$ on the loss function $\Phi$ and because of the similarity to the results obtained in [9] it is reasonable to call a class of probability measures $\mathcal{P}$ "compact" if B is $\sigma(B, L_{\infty,\mathcal{P}}(Y))$-compact. This coincides with the terminology of Pitcher [9] and Rosenberg [11] when $X = R$.

## 3. Existence of minimal complete classes of estimators.

3.1. Theorem. Suppose that the class of considered estimators is given by

$$\mathcal{E} = \{x(\cdot) \in L_{\Phi,\mathcal{P}}(X): R(x(\cdot),P) = \int \Phi(x(t)-g(P))P(dt) < \infty \text{ for each } P \in \mathcal{P}\}.$$

If $\mathcal{P}$ is compact, if the loss function $L(x,P) = \Phi(x - g(P))$, where $\Phi$ is a finite N-function with the complementary function $\Psi$ bounded above on balls centred at zero in Y, then a minimal complete class of estimators in $\mathcal{E}$ exists.

Proof of this Theorem is analogous to that of Pitcher [9; Theorem 3.3] (compare also [5; Theorem 2.4]).

Note that Theorem 3.1 extends Theorem 3.3 of Pitcher onto a larger class of estimators and a larger class of loss functions.

3.2. Corollary to Theorem 3.1. Let $K(P) = \inf_{x(\cdot)\in\mathcal{E}} \sup_{P\in\mathcal{P}} R(x(\cdot),P)$. If $K(P)$ is finite and the assumptions of Theorem 3.1 are fulfilled, then there exists in $\mathcal{E}$ an admissible and minimax estimator of g(P).

The Corollary is an immediate consequence of Theorem 3.1. Note that it improves and extends a theorem of Wertz [13; Satz 1], where norms in Orlicz spaces instead of the risk function were considered.

## References

[1] Fremlin, D.H., Topological Riesz Spaces and Measure Theory. Cambridge University Press 1974.

[2] Kościelska, H., Necessary and sufficient conditions for admissibility. Preprint No 194, Institute of Mathematics Polish Acad. of Sci. October 1979.

[3] Kozek, A., Orlicz spaces of functions with values in Banach spaces. Comment. Math. (Prace Mat.) 19, 259-288 (1976).

[4] Kozek, A., Convex integral functionals on Orlicz spaces. Comment. Math. (Prace Mat.) - to appear, Preprint No 89 (May 1976), Institute of Mathematics, Polish Academy of Sciences.

[5] Kozek, A., On the theory of estimation with convex loss functions. Proceedings of the Symposium to Honour Jerzy Neyman, PWN Polish Scientific Publishers, Warszawa 1977.

[6] Kuratowski, K., Ryll-Nardzewski, C., A general theorem on selectors. Bull. Acad. Pol. Sci. Sér. Sci. Math. Astronom. Phys. 13; 397-403 (1965).

[7] Kusama, T., Yamada, S., On Compactness of the Statistical Structure and Sufficiency. Osaka J. Math. 9, 11-18 (1972).

[8] LeCam, L., Summary of results from decision theory (unpublished Notes, No 258, April 1962).

[9] Pitcher, T.S., A more general property than domination for sets of probability measures. Pacific J. Math. 15, 597-611 (1965).

[10] Rockafellar, R.T., Level sets and continuity of conjugate convex functions. Trans. Amer. Math. Soc. 123, 46-63 (1966).

[11] Rosenberg, R.L., Orlicz spaces based on families of measures. Studia Math. 35, 15-49 (1970).

[12] Wagner, D.H., Survey of measurable selection theorems. SIAM J. Control Opt. 15, 859-903 (1977).

[13] Wertz, W., Invariante und optimale Dichteschätzungen. Math. Balkanica 4.129,707-722 (1974).

# "EINE BEMERKUNG ZUM VERGLEICH VON ZWEISEITIGEN TESTPROBLEMEN"

Norbert Kusolitsch
Wien

## A B S T R A C T

Blackwell [1], [2], Chernoff [4] and Lindley [9] introduced different concepts concerning information contained in an experiment. In this paper a result is proved that resembles at some extent to a theorem of Muirhead (Hardy [6]). By means of this result it is shown that dichotomies which are "more informative" in the sense of Blackwell are also more informative" in the sense of Chernoff or Lindley. The converse is not true.

## E I N L E I T U N G

Seien $(M,\mathfrak{A})$, $(M',\mathfrak{A}')$ Meßräume und $P,Q(P',Q')$ Wahrscheinlichkeitsverteilungen auf $(M,\mathfrak{A})$ (bzw. $(M',\mathfrak{A}')$). Mit $p,q(p',q')$ wollen wir die Dichten von $P,Q(P'Q')$ bezüglich eines dominierenden Maßes $\mu(\mu')$ bezeichnen; wenn nicht anders vermerkt, ist $\mu=\frac{P+Q}{2}(\mu'=\frac{P'+Q'}{2})$.

Definition 1: Ein Quadrupel $T=(P,Q,M,\mathfrak{A})$, bestehend aus einem Meßraum $(M,\mathfrak{A})$ und 2 Verteilungen $P,Q$ auf $(M,\mathfrak{A})$, wird ein Testproblem genannt.

Definition 2: Ist $T=(P,Q,M,\mathfrak{A})$ ein Testproblem, so heißt die Menge $R_T:=\{(\int\varphi dP, \int(1-\varphi)dQ) : \varphi : (M,\mathfrak{A}) \rightarrow ([0,1], \mathfrak{B}_{[0,1]})\}$ Risikomenge von T.

Die untere Grenze der Menge $\{(x,y) : (1-x,y)\in R_T\}$ kann in Parameterform folgendermaßen angeschrieben werden:

$$\{(\alpha(k,\delta),\ r_T(\alpha(k,\delta)),\ k\in[0,\infty],\ \delta\in[0,1]\} \quad \text{mit}$$

$$\alpha(k,\delta) := P[\tfrac{q}{p} < k] + \delta P[\tfrac{q}{p} = k]$$

$$r_T(\alpha(k,\delta)) := Q[\tfrac{q}{p} < k] + \delta Q[\tfrac{q}{p} = k]$$

Anstelle von $r_T$ werden wir mitunter auch $r_{(P,Q)}$ schreiben.

Definition 3: Das Testproblem $T=(P,Q,M,\mathfrak{A})$ ist gemäß Blackwell informativer als ein Testproblem $T'=(P',Q',M',\mathfrak{A}')$ -im Zeichen $T\supset T'$- wenn $R_T \supseteq R_{T'}$, oder anders ausgedrückt, wenn $r_T(\alpha) \le r_{T'}(\alpha)\ \forall\ \alpha\in[0,1]$ (vgl Blackwell [2])

Die Funktion $r_T(\alpha)$ hat folgende Eigenschaften (vgl Österreicher [10], Witting [12]):

r1) $r_T(0) = 0$, $r_T(1)=1$ (oder nicht eindeutig definiert)

r2) $r_T(\alpha)$ ist monoton wachsend und konvex

r3) sei $D_- r_T(\alpha(k,\delta))$ die linksseitige Ableitung von $r_T(\alpha)$, dann ist $k := D_- r_T(\alpha(k,\delta))\ \forall\ k\in[0,\infty], \delta\in[0,1]$

## C H E R N O F F S   K O N Z E P T

Chernoff [4] hat jedem Testproblem $T=(P,Q,M,\mathfrak{A})$ einen Index $\rho := \inf\limits_{0\le\gamma\le 1} \int q^\gamma p^{1-\gamma} d\mu$ zugeordnet. Der nächste Satz zeigt, wo die Bedeutung von $\rho$ liegt.

Satz 1: Ordnet man den beiden Testproblemen $T=(P,Q,M,\mathfrak{A})$ und $T'=(P',Q',M',\mathfrak{A}')$ die Indizes $\rho$ und $\rho'$ zu und bezeichnet man das Bayessche Risiko bei $n(n')$ unabhängigen Versuchsdurchführungen, je nachdem ob $T$ oder $T'$ vorliegt, mit $b_n$ oder $b'_{n'}$ (dh $b_n = \inf\limits_{0\le\lambda\le\infty} \{Q[\prod_{i=1}^{n} \frac{q(x_i)}{p(x_i)} < k] + P[\prod_{i=1}^{n} \frac{q(x_i)}{p(x_i)} \ge k]\}$)

so gilt unabhängig von k:

$$\lim_{n,n'\to\infty} \frac{n'}{n} < \frac{\log \rho}{\log \rho'} \Rightarrow \lim_{n,n\to\infty} \frac{b_n}{b'_{n'}} = \infty$$

$$\lim_{n,n'\to\infty} \frac{n'}{n} > \frac{\log \rho}{\log \rho'} \Rightarrow \lim_{n,n'\to\infty} \frac{b_n}{b'_{n'}} = 0$$

Etwas grob gesprochen, heißt das, daß sich die Bayeschen Risiken asymptotisch annähernd gleich verhalten, wenn die Stichprobenumfänge zueinander ungefähr im Verhältnis $\log \rho/\log \rho'$ stehen. Ist demnach $\rho$ klein verglichen mit $\rho'$, so sind beim Testproblem T' entsprechend mehr Versuche durchzuführen als bei T um die gleiche Fehlerwahrscheinlichkeit zu erhalten (da $\rho,\rho'\in[0,1]$ und deshalb $\log \rho$, $\log \rho'<0$)

Bezeichnet man mit F die Verteilung von p unter $\mu$, also $F(t):= \mu[p(x)\leq t]\ \forall t\in \mathbb{R}$, so gilt

$$F(0) = 0,\ F(2) = 1$$

und $\int\limits_{[0,2]} t\,dF(t) = \int\limits_{\{x:p(x)\in[0,2]\}} p(x)\,d\mu(x) = P(M)=1$

sowie $\int\limits_{[0,2]} (2-t)\,dF(t)=1$

Wegen $p+q\equiv 2$ kann $H_\gamma(P,Q):= \int\limits_M q^\gamma\ p^{1-\gamma} d\mu$ in der Form

$H_\gamma(P,Q) = \int\limits_{[0,2]} (2-t)^\gamma t^{1-\gamma} dF(t)$ angeschrieben werden.

Offensichtlich ist $\psi(t):=(2-t)^\gamma t^{1-\gamma}$ konkav.

Sei nun $L_F$ die Lorenz-Kurve von F, in Parameterform

$$u(\beta):= \int\limits_{[0,\beta]} dF,\quad v(\beta) = \int\limits_{[0,\beta]} t\ dF.$$

Es gilt $u(\beta) = \frac{P+Q}{2}[x:p(x)\leq\beta]$, $v(\beta) = P[x:p(x)\leq\beta]$.

Wegen q=2-p kann auch $(\alpha, r_T(\alpha))$ in der folgenden Weise dargestellt werden:

$$\alpha(\beta,\delta) = P[\beta<p] + \delta P[\beta=p]. \text{ mit } \beta=2/(k+1)$$

$$r_T(\alpha(\beta,\delta)) = Q[\beta<p] + \delta Q[\beta=p]$$

Bezeichnet man mit F' die zu T' gehörige Verteilung von p' und mit $L_{F'}$ die entsprechende Lorenzkurve, so sieht man leicht, daß aus $r_T \leq r_{T'}$ folgt $L_F \leq L_{F'}$. Aus $L_F \leq L_{F'}$ folgt

$$\int_{[0,u]} F^{-1}(s)ds \leq \int_{[0,u]} F'^{-1}(s)ds \quad \forall u\in[0,1]$$

Daraus ergibt sich wegen $\int_{[0,\beta]} F(t)dt = \sup_{u\geq 0}\{\beta u - \int_{[0,u]} F^{-1}(s)ds\}$

und $\int_{[0,\beta]} F(t)dt = \int_{[0,\beta]}(\beta-t)dF(t)$: $\int_{[0,\beta]}(\beta-t)dF(t) \geq \int_{[0,\beta]}(\beta-t)dF'(t)$. (siehe [13])

Da jede konvexe Funktion h(t) auf (0,1) durch Funktionen der Gestalt $c(t) = \begin{cases}\beta-t & \text{für } t\leq\beta \\ 0 & \text{sonst}\end{cases}$ approximiert werden kann, gilt

$$\int h(t)dF(t) \geq \int h(t)dF'(t) \text{ für jedes konvexe h.}$$

Somit

$$H_\gamma(P,Q) \leq H_\gamma(P',Q') \quad \forall\gamma\in(0,1)$$

und es gilt folgender Satz

Satz 2: Gilt für die beiden Testprobleme T und T' mit den zugehörigen Indizes $\rho$ und $\rho'$: $T \supset T'$ so auch

$$H_\gamma(P,Q) \leq H_\gamma(P',Q') \quad \forall\gamma\in(0,1)$$
$$\rho\leq\rho'$$

Die Umkehrung gilt nicht, wie das folgende Beispiel zeigt.

Beispiel: $M=M'=\{1,2,3\}$ $\mathfrak{A}=\mathfrak{A}'=p(M)$

$P=(\frac{1}{2},\frac{1}{2},0),\ Q=(0,\frac{1}{2},\frac{1}{2}),\ P'=(\frac{4}{5},\frac{1}{5},0),\ Q'=(\frac{1}{5},\frac{4}{5},0)$

$$r_1(\alpha)=(\tfrac{1}{2}-\alpha)I_{[0,\frac{1}{2}]}+0\cdot I_{(\frac{1}{2},1]}$$

$$r_2(\alpha)=(1-4\alpha)\,I_{[0,\frac{1}{5}]}+(\tfrac{3}{20}-\tfrac{\alpha}{4})I_{(\frac{1}{5},1]}$$

Es gilt demnach weder $r_1 \le r_2$ noch $r_2 \le r_1$, aber

$$H\gamma(P,Q)=(\tfrac{1}{2})^{\gamma}(\tfrac{1}{2})^{1-\gamma}=\tfrac{1}{2}\quad \forall\gamma\in(0,1)$$

$$H\ (P',Q')=(\tfrac{4}{5})^{\gamma}(\tfrac{1}{5})^{1-\gamma}+(\tfrac{1}{5})^{\gamma}(\tfrac{4}{5})^{1-\gamma}=\tfrac{1}{5}[4^{\gamma}+4^{1-\gamma}]\ge\tfrac{4}{5}\ge\tfrac{1}{2}\quad \forall\gamma\in(0,1)$$

und $\rho \le \rho'$.

## L I N D L E Y S   K O N Z E P T

Im Gegensatz zu Blackwell und Chernoff geht Lindley [9] davon aus, daß der Statistiker, wenn er ein Testproblem zu untersuchen hat, bereits ein gewisses Wissen über die Häufigkeit des Auftretens von P oder Q besitzt, welches durch eine a-priori Verteilung $\overline{\pi}=(\pi,1-\pi)$ $\pi\in[0,1]$ über $\{P,Q\}$ beschrieben wird. Um nun die durch eine Beobachtung x gewonnene Information über T zu beschreiben, folgt Lindley dem von Shannon vorgezeichneten Weg.

Definition 4: Sei $T=(P,Q,M,\mathfrak{A})$ ein Testproblem, $\overline{\pi}=(\pi,1-\pi)$ eine a-priori Verteilung und x eine Beobachtung.
Mit den Bezeichnungen

$$p(0|x):=\frac{\pi p(x)}{\pi p(x)+(1-\pi)q(x)}\ ,\quad p(1|x):=\frac{(1-\pi)q(x)}{\pi p(x)+(1-\pi)q(x)}$$

heißt $I(T,\overline{\pi},x):=p(0|x)\log p(0|x)+p(1|x)\log p(1|x)-\pi\log\pi - -(1-\pi)\log(1-\pi)$

die durch die Beobachtung x bei Vorliegen der a-priori Verteilung $\bar{\pi}$ gewonnene Information über T.

Definition 5: $I(T,\bar{\pi}):=\int I(T,\bar{\pi},x)[\pi p(x)+(1-\pi)q(x)]d\mu(x)$ ist die durchschnittliche Information von T bei a-priori Verteilung $\bar{\pi}$.

$$I(T,\bar{\pi})=-\pi\int p(x)\log\frac{\pi p(x)+(1-\pi)q(x)}{p(x)}d\mu(x)-(1-\pi)\int q(x)\log\frac{\pi p(x)+(1-\pi)q(x)}{q(x)}d\mu(x)=$$

$$=I(P|\pi P+(1-\pi)Q)+I(Q|\pi P+(1-\pi)Q)$$

Dabei bezeichnet man mit $I(P|Q)$ die I-Divergenz $\int p\log\frac{p}{q}d\mu$ zwischen P und Q.

Falls P und Q äquivalent sind hängen I-Divergenz und Divergenzen der Ordnung $\gamma$ $H_\gamma(P,Q)$ folgendermaßen zusammen (vgl Vajda [11])

$$\lim_{\gamma\to 1}\frac{1}{\gamma-1}\log H_\gamma(P,Q)=I(P|Q)$$

Unmittelbar aus Satz 2 ergibt sich dann

Korollar 1: Sind $T=(P,Q,M,\mathfrak{A})$, $T'=(P',Q',M',\mathfrak{A}')$ 2 Testprobleme, wobei $T\supset T'$ gilt und P und Q äquivalent sind, so ist $I(P|Q)$ größer oder gleich zu $I(P'|Q')$

Übrigens gilt das Korollar auch dann, wenn P und Q nicht äquivalent sind.

Definition 5: T ist im Sinne von Lindley informativer als T', wenn $I(T,\bar{\pi})\geq I(T',\bar{\pi})$ $\forall\bar{\pi}=(\pi,1-\pi)$, $\pi\in[0,1]$.

Im wesentlichen ist $r_{(P,\pi P+(1-\pi)Q)}$ durch die Punkte

$(P[\frac{\pi p+(1-\pi)q}{p}\leq k]$, $(\pi P+(1-\pi)Q)[\frac{\pi p+(1-\pi)q}{p}\leq k]$ $k\in[\pi,\infty]$ bestimmt.

Da $P[\frac{\pi p+(1-\pi)q}{p}\leq k]$ bzw. $(\pi P+(1-\pi)Q)[\frac{\pi p+(1-\pi)q}{p}\leq k]$ auf die

Form $P[\frac{q}{p}\leq k':=\frac{k-\pi}{1-\pi}]$ bzw. $(\pi P+(1-\pi)Q)[\frac{q}{p}\leq k']$ gebracht werden können,

folgt aus $r_T\leq r_T'$, daß zu jedem $k\in[\pi,\infty]$ ein $k^*\geq\pi$ und ein $\delta\in[0,1]$ existieren, sodaß

$$P[\frac{\pi p+(1-\pi)q}{p}\le k]=P'[\frac{\pi p'+(1-\pi)q'}{p'}<k^*] + \delta P'[\frac{\pi p'+(1-\pi)q'}{p'} = k^*]$$

$$Q[\frac{\pi p+(1-\pi)q}{p}\le k]\le Q'[\frac{\pi p'+(1-\pi)q'}{p'}<k^*] + \delta Q'[\frac{\pi p'+(1-\pi)q'}{p'} = k^*]$$

Demnach folgt aus $r_T\le r_{T'}$ auch $r_{(P,\pi P+(1-\pi)Q)}\le r_{(P',\pi P'+(1-\pi)Q')}$

was wieder zur Folge hat, daß $I(P|\pi P+(1-\pi)Q)\ge I(P'|\pi P'+(1-\pi)Q')$.

In analoger Weise zeigt man $I(Q|\pi P+(1-\pi)Q)\ge I(Q'|\pi P'+(1-\pi)Q')$
Es gilt also

Satz 3: Ist $T\supset T'$, so ist T auch informativer im Sinne von Lindley als T'.

Dieser Satz wurde bereits von Lindley [9] gezeigt, allerdings verwendet er anstelle von $T\supset T'$ die Voraussetzung $T\succ T'$. Dabei bedeutet $T\succ T'$, daß eine stochastische Transformation W existiert, sodaß $WP=P'$ und $WQ=Q'$. Blackwell [2] hat gezeigt, daß die Relationen "$\supset$" und "$\succ$" äquivalent sind; in [2] finden sich auch Kriterien betreffend die Relation "$\succ$". Für weitere Details betreffend Lindleys Konzept sei vor allem auf die Arbeit von Bussgang-Marcus [3] verwiesen. Einen guten Überblick über verschiedene Methoden zum Vergleich von Experimenten findet man bei Nemetz [8].

Abschließend möchte ich Prof. T.Nemetz und Dr. F.Österreicher für viele wertvolle Hinweise und Ratschläge danken. Dem unbekannten Begutachter sei an dieser Stelle für die wesentlich vereinfachte Fassung des Beweises von Satz 2 gedankt.

## L I T E R A T U R

[1] Blackwell D.(1951): Comparison of experiments. Proceedings of the 2nd Berkeley Symposium on Mathematical Statistics and Probability, University of California Press, 93-102

[2] Blackwell D.(1953): Equivalent comparisons of experiments. Ann. Math. Stat., Vol 24, 265-272

[3] Bussgang J.J. - Marcus M.B.(1967): Information Theory and Alternate Hypothesis Tests, Journal of Optimization Theory and Applications, Vol. 1, No 3, 194-214

[4] Chernoff H.(1952): A measure of asymtotic efficiency for tests of a hypothesis based on the sum of observations. Ann. Math. Stat. 23, 493-507

[5] Csiszar I. (1969): On Generalized Entropy. Stad.Sci. Math.Hung. 4, 401-419

[6] Hardy G. - Littlewood J.E. - Pôlya G. (1952): Inequalities.Cambridge University Press

[7] Hewitt E. - Stromberg K. (1965): Real and Abstract Analysis. Springer, Berlin - New York

[8] Nemetz T. (1977): Information Type Measures and Their Applications to Finite Decision Problems. Lectures Notes Carleton University, Ottawa

[9] Lindley D.V. (1956): On a Measure of the Information Provided by an Experiment. Ann Math. Stat. Vol 27, No 4, 986-1005

[10] Österreicher F. (1972): An Information - type Measure of Difference of Probability Distributions Based on Testing Statistical Hypotheses. Coll. Math. Soc. Jânos Bolyai, 593-600

[11] Vajda I. (1970): On the Amount of the Information Contained in a Sequence of Independent Observations, Kybernetika 6, 306-323

[12] Witting H. (1966): Mathematische Statistik, Teubner, Stuttgart

[13] Krasnosel´skii M.A. - Rutickii Ya.B. (1961): Convex Functions and Orlicz Spaces, Noordhoff, Groningen

Duality of the maximal inequality for non-negative submartingales and of the convexity inequality of Burkholder

J.Mogyoródi
University of Budapest

1. In this paper we shall deal with maximal inequalities of the form

/1/ $$E(\Phi(X_n^*)) \leq a + E(\Phi(bX_n)),$$

where $(X_n, \mathcal{F}_n)$, $n \geq 1$, is any non-negative submartingale with the maximal function

$$X_n^* = \max_{1 \leq k \leq n} X_k$$

defined on the probability space $(\Omega, \mathcal{A}, P)$; $\Phi$ is a Young-function and $a \geq 0$ as well as $b > 0$ are constants depending only on $\Phi$.

The above maximal inequality is valid in the sense that both sides are at the same time infinite or finite and if they are finite then /1/ holds.

Maximal inequalities of form /1/ are closely related to the so called convexity inequalities. Let $z_1, z_2, \ldots, z_n$, $n \geq 1$, be non-negative random variables with finite expectation and let $\mathcal{F}_1 \subset \mathcal{F}_2 \subset \ldots \subset \mathcal{F}_n$, $n \geq 1$, be an increasing sequence of arbitrary $\sigma$ - fields of events. Consider the random variables

$$Z_n = \sum_{i=1}^{n} z_i$$

and

$$A_n = \sum_{i=1}^{n} E(z_i \mid \mathcal{F}_i), \quad n \geq 1.$$

We say that for the Young-function $\Phi$ the convexity inequality is valid if for arbitrary $n \geq 1$, arbitrary non-negative random variables $z_1, \ldots, z_n$ of finite expectation and for arbitrary increasing sequence of $\sigma$ - fields of events $\mathcal{F}_1 \subset \ldots \subset \mathcal{F}_n$ with the corresponding $Z_n$ and $A_n$ the inequality

$$E(\Phi(A_n)) \leq c + E(\Phi(d Z_n)) \tag{/2/}$$

holds, where $c \geq 0$ and $d > 0$ are constants depending only on $\Phi$.

This inequality is valid in the sense that both sides are at the same time infinite or finite and if they are finite then /2/ holds.

Consider the pair of conjugate Young-functions $(\Phi, \Psi)$. The duality of the maximal inequality and of the convexity inequality is formulated in the following

Theorem 1. For the Young-function $\Phi$ the maximal inequality /1/ holds for any $n \geq 1$ and for all non-negative submartingale $(X_n, \mathcal{F}_n)$ if and only if the convexity inequality /2/ is valid for the conjugate Young-function $\Psi$ . In this case the corresponding constants are the same in /1/ and in /2/.

On the basis of this duality principle we see that when the class of the Young-functions for which one of the above inequalities can be extended to a larger one, then the same is true for the class of the conjugate Young-functions for the other inequality.

2. Maximal inequalities of type /1/ are well-known

in case of the Young-function

$$\Phi(x) = x^p/p \qquad (x \geq 0,\ 1 < p < +\infty).$$

In this case /1/ reduces to Doob's inequality

$$E(X_n^{*p}) \leq q^p E(X_n^p); \quad p^{-1} + q^{-1} = 1$$

A generalization of this is given by Garsia [1]:
We say that the Young-function $\Psi$ satisfies the growth condition if there exist constants $a > 1$ and $A > 0$ such that for all $x \geq 0$ the inequality

$$\Psi(ax) \leq A\Psi(x)$$

holds. This is equivalent to the condition that the quantity

$$q = \sup_{x>0} \frac{x\psi(x)}{\Psi(x)}$$

is finite. Here $\psi(x)$ denotes the right-derivative of $\Psi(x)$. The quantity

$$q = \sup_{x>0} \frac{x\psi(x)}{\Psi(x)}$$

is called the power of $\Psi$. Garsia has shown that if the power $q$ of $\Psi$ is finite then

$$E(\Phi(X_n^*)) \leq qE(\phi(qX_n)),$$

where $\Phi$ is the conjugate of $\Psi$. In [2] we have sharpened this inequality. Namely, we have

Theorem 2. If the power $q$ of $\Psi$ is finite then for all $n \geq 1$ and for arbitrary non-negative submartingale $(X_n, \mathcal{F}_n)$ we have

$$/3/ \qquad E(\phi(X_n^*)) \leq E(\phi(qX_n)).$$

This is the direct generalization of Doob's maximal inequality.

When the power

$$p = \sup_{x>0} \frac{x\varphi(x)}{\Phi(x)}$$

of $\Phi$ is finite and we do not suppose the finiteness

of the power of $\Psi$ then, in general, the maximal inequality /1/ does not hold for $\Phi$. /Here, $\varphi(x)$ denotes the right-derivative of $\Phi$./ The case of the Young-function

$$\Phi(x) = (x+1)\log(x+1) - x$$

shows this. Namely, it is classical that for any non-negative submartingale $(X_n, \mathcal{F}_n)$ we have

$$E(X_n^*) \leq \frac{e}{e-1}\left[1 + E((X_n+1)\log(X_n+1))\right] - 1$$

and this inequality cannot be improved. It has been shown by Gundy [3] that for a sufficiently large class of non-negative martingales the reverse inequality

$$E((X_n+1)\log(X_n+1)) \leq E((X_1+1)\log(X_1+1)) + C\,E(X_n^*) + C$$

is true provided that

$$E((X_1+1)\log(X_1+1))$$

is finite. Here $C$ is the constant characteristic to the considered martingale.

However, there is a sufficiently large class of Young-functions $\Phi$ with finite power, say $p$, for which a maximal inequality of form /1/ is valid. We have

Theorem 3. Suppose the power $p$ of the Young-function $\Phi$ is finite and

$$\sup_{x \geq 1} \frac{1}{\varphi(x)} \int_1^x \frac{\varphi(t)}{t}\,dt = c < +\infty \qquad /4/$$

holds. Then for all $n \geq 1$ and for any non-negative submartingale $(X_n, \mathcal{F}_n)$ we have

$$E(\Phi(X_n^*)) \leq (p+1)\,\Phi(1) + E(\Phi(pcX_n)).$$

Remark. T.Móri and the author have shown that /4/

is also necessary for the validity of /1/ if $p$ is finite.

For example, when $\Phi(x) = x^p/p$, with power $1 < p < +\infty$, then the above conditions trivially hold.

3. By the duality principle of Theorem 1 the assertions of section 2, concerning the validity of maximal inequalities, enable us to formulate some convexity inequalities. The first of them is due to Burkholder, Davis and Gundy [4]. Garsia has proved this with a better constant. Our method of proof is based on the above duality principle and the constant is also new.

Theorem 4. Let $\Phi$ be a Young-function with finite power $p$. Then the convexity inequality

$$E(\Phi(A_n)) \leq E(\Phi(pZ_n))$$

holds in the sense as it has been defined in section 1.

To prove this, note that on the basis of our assumption by Theorem 2 the maximal inequality /3/ is valid with $\Psi$ and $p$ instead of $\Phi$ and $q$. So, by Theorem 1 the convexity inequality holds for $\Phi$ with the constants $c = 0$ and $d = p$.

One proves similarly the following assertion:

Theorem 5. Let $\Psi$ be a Young-function with finite power $q$.

Then the convexity inequality

$$E(\Phi(A_n)) \leq (q+1)\Psi(1) + E(\Phi(qcZ_n))$$

holds if and only if

$$\sup_{x \geq 1} \frac{1}{\psi(x)} \int_1^x \frac{\psi(t)}{t}\,dt = c < +\infty .$$

4. In the preceding section we deduced convexity inequalities via maximal inequalities. Now by means of another type of convexity inequalities we can deduce a maximal inequality.

It is known that when the random variables $z_i \geq 0$ are such that

$$Z_\infty = \sum_{i=1}^{\infty} z_i \leq K < +\infty$$

then for arbitrary $t \in (0, K^{-1})$ we have

$$E(\exp(tA_\infty)) \leq (1 - tK)^{-1}$$

where

$$A_\infty = \sum_{i=1}^{\infty} E(z_i \mid \mathcal{F}_i)$$

/cf.e.g. [ 5 ], Theorem 46 ,p.49/. From this one easily deduces the following assertion:

Lemma 1. Let $\Psi(x)$ be a Young-function such that for some $t \in (0, K^{-1})$ the integral

$$I = \int_0^{+\infty} e^{-t\lambda} \psi(\lambda)\, d\lambda$$

converges. Then

$$E(\Psi(A_\infty)) \leq (1 - tK)^{-1} I .$$

Using these trivial remarks we can prove the following

Theorem 6. Let $(\Phi, \Psi)$ be a pair of conjugate Young-functions and let $g(x)$ , $x \geq 0$ , be a convex, non-negative and increasing function.
If for some $t \in (0,1)$ the integral

$$I = \int_0^{+\infty} e^{-t\lambda} \psi(\lambda)\, d\lambda$$

converges then for arbitrary non-negative submartingale $(X_n, \mathcal{F}_n)$ we have

$$E(g(X_n^*)) \leq E(\Phi(g(X_n))) + (1-t)^{-1} I ,$$

provided that

$$E(\Phi(g(X_n)) < +\infty .$$

Put in this inequality $g(x) \equiv x$ , $x \geq 0$ .
Then under the other conditions of this theorem we have

$$E(X_n^*) \leq E(\Phi(X_n)) + (1-t)^{-1} I,$$

provided that

$$E(\phi(X_n)) < +\infty .$$

Especially, with

$$\varphi(x) = d \log(1+x),$$

where $d > 2$ is a constant, we have

$$\psi(\lambda) = \exp\left(\frac{\lambda}{d}\right) - 1 .$$

Consequently, choosing $t = 1/2$ and $d = 4$ we get that

$$I = \int_0^{+\infty} \exp\left(-\frac{\lambda}{2}\right)\left(\exp\left(\frac{\lambda}{4}\right) - 1\right) d\lambda = 2$$

and that

$$E(X_n^*) \leq 4 \left[ E((X_n+1) \log(X_n+1) + 1 \right] .$$

So, Theorem 6 contains the classical Doob inequality for the non-begative submartingales belonging to L log L.

In this connection Gundy [3] has shown that for a large class of non-negative martingales the last inequality cannot be improved. Let $(X_n, \mathcal{F}_n)$ be a non-negative martingale satisfying the Gundy condition: for all $n \geq 1$ the inequality

$$X_{n+1} \leq C X_n \qquad /5/$$

holds a.e., where $C > 0$ is a constant. Then

$$E((X_n+1) \log(X_n+1)) \leq E((X_1+1) \log(X_1+1)) + C E(X_n^*+1),$$

provided that $E((X_1+1) \log(X_1+1)) < +\infty$ .

In the following assertion we describe those Young-functions for which the inequality of the preceding theorem, i.e. the inequality

$$E(X_n^*) \leq E(\Phi(X_n)) + (1-t)^{-1} I$$

can be reversed.

Theorem 7. Let $\Phi$ be a Young-function such that

$$x\varphi(x) - \Phi(x) = \mathcal{O}(x), \quad x \to +\infty.$$

Then for any non-negative martingale $(X_n, \mathcal{F}_n)$ satisfying the condition /7/ of Gundy we have

$$E(\Phi(X_n)) \leq K\left[E(\Phi(X_1)) + E(X_n^*) + 1\right],$$

provided that $E(\Phi(X_1)) < +\infty$. Here $K > 0$ is a constant depending only on $\Phi$.

For example in case of the Young-function

$$\Phi(x) = (x+1)\log(x+1) - x$$

we have

$$x\varphi(x) - \Phi(x) = x - \log(1+x) \leq x$$

for arbitrary $x \geq 0$. So, in this case the condition of the preceding theorem is satisfied.

One can show that this condition is in some sense necessary, too. All these assertions are treated in detail in [6] and [7].

References

[1] Garsia, A.M.: A convex function inequality for martingales. The Annals of probability. 1/1973/ 171-174.

[2] Mogyoródi, J.: On an inequality of Marcinkiewicz and Zygmund. Publicationes Mathematicae. Debrecen. 26/1979/ 267-274.

[3] Gundy,R.F.: On the class L log L, martingales and singular integrals. Atudia Mathematica. 33/1969/ 109-118.

[4] Burkholder,D.L. , Davis,B.J. and Gundy,R.F.: Integral inequalities for convex functions of operators on martingales. Proceedings 6-th. Berkeley Symp.on Math. Stat.and Probability.University of California Press. 1972. 223-240.

[5] Meyer,P.A.: Martingales and Stochastic integrals I.Lecture Notes in Mathematics.Springer, Berlin.284.1972.

[6] Mogyoródi,J.: Maximal inequalities, convexity inequality and their duality.I. Submitted to Analysis Mathematica.

[7] Mogyoródi,J.: Maximal inequalities, convexity inequality and their duality.II. Submitted to Analysis Mathematica.

# ON A HOEFFDING-TYPE PROBLEM

by

Tamás F. Móri

L. Eötvös University, Budapest

It is well-known from a general theorem of Hoeffding [2] that the limit distribution of a sequence of the so-called U-statistics is normal. However, the variance of this normal distribution frequently vanishes as a consequence of improper norming. Hoeffding's theorem is not applicable even if the statistics are elementary symmetric polynomials of independent identically distributed random variables. In fact, the limit distributions are not normal in these cases.

In this paper we prove that the limit distribution of polynomial U-statistics (under natural conditions on the existence of some moments) can be expressed as a polynomial of multidimensional normal distribution.

Let us introduce some definitions following Hoeffding's paper [2]. Let $X_1, X_2, \ldots$ be i.i.d. random variables and let

$$\Phi=\Phi(x_1,x_2, \ldots ,x_k)$$

be a symmetric function in $k$ variables. Consider the U-statistics of $X_1, X_2, \ldots , X_n$ constructed by the help of $\Phi$:

$$U_n = \binom{n}{k}^{-1} \sum_{1\leq i_1<i_2< \ldots <i_k\leq n} \Phi(X_{i_1},X_{i_2}, \ldots ,X_{i_k}) \quad .$$

For $d=1, 2, \ldots , k$ let

$$\Phi_d(x_1, \ldots ,x_d) = \mathrm{E}\Phi(x_1, \ldots ,x_d,X_{d+1}, \ldots ,X_k) \; .$$

For the sake of brevity let us write $E_d(.)$ instead of the conditional expectation $E(. \mid X_1, \dots, X_d)$. The equality

$$\phi_d(X_1, \dots, X_d) = E_d\phi(X_1, \dots, X_k) \tag{1}$$

implies that

$$E\phi_d(X_1, \dots, X_d) = EU_n \quad .$$

Without loss of generality we can suppose that this common expectation equals zero. Denote the variance of $\phi_d(X_1, \dots, X_d)$ by $\zeta_d$ . In order to assure the finiteness of all $\zeta_d$ it is sufficient to require that $\zeta_k$ be finite; this latter will be assumed throughout our work. Since

$$\begin{aligned} &E\big(\phi(X_1, \dots, X_d, X_{d+1}, \dots, X_k)\phi(X_1, \dots, X_d, X_{k+1}, \dots, X_{2k-d})\big) = \\ &= E\big[E_d(\phi(X_1, \dots, X_d, X_{d+1}, \dots, X_k)\phi(X_1, \dots, X_d, X_{k+1}, \dots, X_{2k-d}))\big] = \\ &= E\big[E_d\phi(X_1, \dots, X_d, X_{d+1}, \dots, X_k)\, E_d\phi(X_1, \dots, X_d, X_{k+1}, \dots, X_{2k-d})\big] = \\ &= E\phi_d^2(X_1, \dots, X_d) = \zeta_d \ , \end{aligned}$$

therefore the variance of $U_n$ is

$$EU_n^2 = \binom{n}{k}^{-1} \sum_{d=1}^{k} \binom{k}{d}\binom{n-k}{k-d} \zeta_d \ .$$

It is obvious from (1) that if $\zeta_d=0$ then $\zeta_1=\zeta_2= \dots =\zeta_{d-1}$ as well. Suppose that $\zeta_d$ is the first non-zero element of the sequence $\zeta_1, \zeta_2, \dots , \zeta_k$. In this case we say $\phi$ to be a $d$-type function with respect to the distribution of $X_i$ . (Here we depart from the terminology of Hoeffding [2]: having regarded $E\phi(X_1, \dots, X_k)$ as a functional of the distribution function of the random variables $X_i$, Hoeffding called this functional stationary of order $d-1$.)

The following theorem shows that we can set $d=k$ without loss of generality.

THEOREM 1.

*Denote* $Z_n = \binom{n}{d}^{1/2}\binom{k}{d}^{-1} U_n$ *and* $Y_n = \binom{n}{d}^{-1/2} \sum_{1 \le i_1 < \dots < i_d \le n} \phi_d(X_{i_1}, \dots, X_{i_d})$.

*Then*

$$E(Z_n - Y_n)^2 = O(1/n).$$

From (1) one can see that $\phi_d$ is of $d$-type with respect to the distribution of $X_i$.

*Proof*. The proof is like that of Theorem 7.1. of [2].

$$E(Z_n - Y_n)^2 = EZ_n^2 + EY_n^2 - 2\, EZ_nY_n$$

Here

$$EZ_n^2 = \binom{n}{d}\binom{k}{d}^{-2} EU_n^2 = \binom{n}{k}^{-1}\binom{k}{d}^{-1}\binom{n-k}{k-d}\binom{n}{d} \zeta_d + O(1/n) = \zeta_d + O(1/n)$$

moreover

$$E\left(\phi_d(X_1, \dots, X_j, X_{j+1}, \dots, X_d)\phi_d(X_1, \dots, X_j, X_{d+1}, \dots, X_{2d-j})\right) =$$

$$= E\left[E_j\left(\phi_d(X_1, \dots, X_j, X_{j+1}, \dots, X_d)\phi_d(X_1, \dots, X_j, X_{d+1}, \dots, X_{2d-j})\right)\right] =$$

$$= E\left[E_j\phi_d(X_1, \dots, X_j, X_{j+1}, \dots, X_d)E_j\phi_d(X_1, \dots, X_j, X_{j+1}, \dots, X_{2d-j})\right] =$$

$$= E\phi_j^2(X_1, \dots, X_j) = \zeta_j$$

and similarly

$$E\left[\phi(X_1, \dots, X_j, X_{j+1}, \dots, X_k)\phi_d(X_1, \dots, X_j, X_{k+1}, \dots, X_{k+d-j})\right] = \zeta_j$$

thus we have

$$EY_n^2 = \sum_{j=1}^{d}\binom{d}{j}\binom{n-d}{d-j}\zeta_j = \zeta_d$$

and finally

$$EZ_nY_n = \binom{k}{d}^{-1}\sum_{j=1}^{d}\binom{k}{j}\binom{n-k}{d-j}\zeta_j = \zeta_d\,.$$

This completes the proof.

In the following we shall deal with polynomial U-statistics, i.e. we suppose that $\phi$ is a polynomial (then $\phi_d$ is a polynomial, too). According to the preceding theorem we assume

$$\phi_{k-1}(X_1, \dots, X_{k-1}) = E_{k-1}\phi(X_1, \dots, X_k) = 0 \qquad \text{a.s.} \tag{2}$$

What kind of $k$-variate polynomials has this property? As an obvious example one can find the product $x_1 x_2 \dots x_k$, if only the expectation of the random variables $X_i$ is $0$. In this case $U_n$ is just the elementary symmetric polynomial of order $k$ in $X_1, \dots, X_n$. Let us determine the limit distribution in this simple case.

Denote $\{G_k: k=0,1,2, \dots \}$ the orthonormal system of polynomials with respect to the standard normal distribution, i.e. let $G_k = H_k/\sqrt{k!}$ where $H_k(x) = (-1)^k e^{x^2/2} \frac{d^k}{dx^k} e^{-x^2/2}$ is the Hermite polynomial of degree $k$.

Let $N$ be a random variable with standard normal distribution.

THEOREM 2.

*Let $X_1, X_2, \dots$ be i.i.d. random variables with mean 0 and variance 1 (no other moment condition is supposed). Denote*

$$S_n(0)=1, \qquad S_n(k) = \sum_{1 \le i_1 < \dots < i_k \le n} X_{i_1} X_{i_2} \dots X_{i_k} \qquad k=1,2, \dots ,n$$

*the elementary symmetric polynomial of order $k$ in $X_1, \dots, X_n$. Then the limit distribution as $n \to \infty$ of $\binom{n}{k}^{-1/2} S_n(k)$ is the distribution of $G_k(N)$.*

*Proof.* As a starting point of the proof one can use Newton's identities on the connection between elementary symmetric polynomials and power sums.

Denote $Q_n(k) = \sum_{j=1}^{n} X_j^k$. If $n \ge k$ then by Newton's identities

$$Q_n(k)S_n(0) - Q_n(k-1)S_n(1) + \dots + (-1)^{k-1} Q_n(1) S_n(k-1) + (-1)^k k S_n(k) = 0 . \tag{3}$$

Hence $S_n(k)$ can be expressed as a polynomial $P_k$ of the power sums $Q_n(1), Q_n(2), \dots, Q_n(k)$. By the strong law of large numbers

$$\lim_{n \to \infty} n^{-1} Q_n(2) = 1 \qquad \text{a.s.}$$

Hence

$$\lim_{n \to \infty} n^{-1/2} \max_{1 \le j \le n} |X_j| = 0 \qquad \text{a.s.,}$$

and for $k>2$ we have

$$\lim_{n\to\infty} n^{-k/2} Q_n(k) = 0 \qquad \text{a.s.}$$

From this we obtain that

$$\binom{n}{k}^{-1/2} S_n(k) \sim \sqrt{k!}\, P_k\big(n^{-1/2} Q_n(1),\, 1,\, 0,\, \ldots,\, 0\big) \quad \text{a.s.} \tag{4}$$

letting $n\to\infty$.

Denote $H_k(x) = k!\, P_k(x,1,0,\ldots,0)$. Then $H_0=1$, $H_1=x$ and (3) implies

$$H_k = x\, H_{k-1} - (k-1) H_{k-2} \qquad k=2,\, 3,\, \ldots$$

Therefore $H_k$ is the Hermite polynomial of degree $k$. Now we can finish the proof combining (4) with the central limit theorem.

The case of elementary symmetric polynomials is not so special as it seems to be. Namely, the following characterization is valid.

THEOREM 3.

*Let $\phi(x_1, \ldots, x_k)$ be a k-variate symmetric polynomial of k-type with respect to the distribution of $X_i$. Then there exist some constants $\alpha_j$ and (univariate) polynomials $P_j(x)$ for which $EP_j(X_i)=0$ and*

$$\phi(X_1, \ldots, X_k) = \sum_j \alpha_j \prod_{i=1}^{k} P_j(X_i) \qquad a.s. \tag{5}$$

*Moreover the degree of the polynomials $P_j$ does not exceed that of $\phi$ in one argument.*

*Proof.* Disregarding first the $k$-type property we prove the following statement: Any symmetric polynomial has a representation of the form (5) (without any restriction on $EP_j(X_i)$ of course).

In order to prove this latter it is sufficient to show that it holds for elementary symmetric polynomials, for any symmetric polynomial can be expressed as a polynomial of the elementary symmetric functions.

Let $c_0, c_1, \ldots, c_k$ be different real numbers. Let us try to give the constants $\alpha_0, \alpha_1, \ldots, \alpha_k$ such that

$$S_k(l)=\sum_{j=0}^{k}\alpha_j\prod_{i=0}^{k}(x_i+c_j) \qquad 1\leq l\leq k$$

From this the following system of equations can be obtained for $\alpha_0, \alpha_1, \ldots, \alpha_k$

$$\sum_{j=0}^{k} c_j^{k-m}\alpha_j = \delta_{lm} \qquad m=0, 1, \ldots, k.$$

This system of equations is solvable, because the determinnant of the system, being a Vandermonde one, does not vanish.

One can readily verify that constructing the representation (5) in the way described above, the polynomials $P_j$ will satisfy the condition imposed on their degree.

Let us pass over to the $k$-type case. Change $P_j(X_i)$ to $P'_j(X_i)= P_j(X_i)-\mathrm{E}P_j(X_i)$ in the above representation and consider the polynomial

$$R(x_1, \ldots, x_k)=\phi(x_1, \ldots, x_k)-\sum_j \alpha_j\prod_{i=1}^{k}P'_j(x_i) \quad .$$

This consists of terms in which the number of variables is less than $k$. Denote $R_k(x_1, \ldots, x_{k-1})$ the sum of all the summands not containing $x_k$. Since $R$ is obviously of $k$-type, therefore

$$\mathrm{E}_{k-1}R(X_1, \ldots, X_k)=0 \qquad \text{a.s.}$$

From this we obtain that $R_k(X_1, \ldots, X_{k-1})$ is equal (with probability one) to a sum of such terms which have at most $k-2$ variables.

Repeating this procedure with $X_j$ in place of $X_k$ $(j=1, \ldots, k-1)$, we find $R(X_1, \ldots, X_k)$ a.s. equal to a sum of terms in which the number of variables does not exceed $k-2$.

Iterating the above line of reasoning after all we have $R(X_1, \ldots, X_k)=0$ a.s. and the proof is completed. (It does not mean that $R(x_1, \ldots, x_k)=0$ algebraically.)

As an immediate consequence of our theorems we have the following

COROLLARY.

*Let $X_1, X_2, \ldots$ be i.i.d. random variables and let $\phi$ be a k-variate symmetric polynomial of d-type with respect to the distribution of $X_i$. Denote m the degree of $\phi$ in any of its arguments. Suppose $EX_i^{2m}<\infty$ and consider the U-statistics $U_n$ formed by the help of $\phi$.*

*Then the limit distribution of $n^{d/2}(U_n-EU_n)$ is the distribution of a finite sum $\sum_j \alpha_j H_d(N_j)$, where $\alpha_j$'s are real constants, the joint distribution the random variables $N_j$ is multivariate normal and $H_d$ stands for the Hermite polynomial of degree d.*

Instead of proving we only add that the distribution of each $N_j$ is standard normal and from the asymptotic equality (4)

$$EN_jN_l=\left(EP_j^2(X_i)EP_l^2(X_i)\right)^{-1/2} EP_j(X_i)P_l(X_i) .$$

*Remarks.*

1. The above limit theorems can easily be generalized to obtain invariance principles.
2. Another interesting (and much more difficult) problem is to describe the asymptotic behaviour of U-statistics in that case, when the kernel $\phi$ varies with $n$ under some rule, e.g. Halász and Székely in [1], Móri and Székely in [3] deal with the elementary symmetric polynomials $S_n(k)$ when $k$ is not fixed but the quotient $k/n$ tends to a limit $c$. In my lecture at PSMS I mentioned these results too, but here I don't want to go into such details.

REFERENCES

[1] HALÁSZ, G. and SZÉKELY, G. J. (1976). On the elementary symmetric polynomials of independent random variables. *Acta Math. Acad. Sci. Hungar.* 28 397-400

[2] HOEFFDING, W. (1948). A class of statistics with asymptotically normal distribution. *Ann. Math. Statist.* 19 293-325

[3] MÓRI, T. F. and SZÉKELY, G. J. (1979). Asymptotic behaviour of symmetric polynomial statistics. *To appear in Ann. Prob.*

# RUN-TEST DISCRIMINATION BETWEEN WRITTEN HUNGARIAN AND RANDOM SEQUENCES

TIBOR NEMETZ

MATH.INST. OF THE HUNGARIAN ACADEMY OF THE SCIENCES

SUMMARY:

The paper deals with a reliable error-detection method in transmitting written Hungarian. It is shown that there is a class of rate-one linear block codes which yields asymptotically vanishing decision-error probabilities. The error-detection goes via testing simple vs simple hypotheses. Theoretical results are combined with results of practical interest: an application of the run-tests allows an acceptable error-detection with block-length 40.

## INTRODUCTION

The reliability of the information transmission is usually ensured by using error-correcting or error-detecting codes. These codes introduce extra redundancy into the messages, i.e. they increase the message-length. The increase is hardly justifiable in the case when the message itself is very redundant, like any written European language. In such cases one can apply invertible random scrambling to blocks of the message as error-detection.

The scrambled text is sent over the noisy channel and the receiver applies the inverse transformation. If the transmission was error-free he gets the original text (which is unknown to him), otherwise he gets a complete mess. It means that the decoded message is distributed either according to the probability law of the source (which is known to the receiver) or it forms a sequence of essentially independent, uniformly distributed random variables. Therefore the error-detection is carried out by testing a simple hypothesis vs. another simple one. If the receiver knows much less abouc the source he still can do reliable decision. E.g. if the source is a stationary one with known non-zero redundancy the receiver can detect any transmission-error with decision-error probabilities going to zero exponentially as the block-length goes to infinity. An obvious statistics to be used for detection in this case in the empirical redundancy of the inverted received text, which converges stochastically to the known value resp. to zero.

The possibility of utilizing the natural redundancy of a source for error detection was first observed by M.Hellman [2] . He used a convolutional encoder, feedback decoder pair to explain the idea. In what follows we apply a similar encoder which allows a straitforward theoretical treatment and the use of easily implementable run-tests in the hypothesis testing.

Random scrambling.

Let $p$ be a given integer, and suppose that a source produces a finite or infinite string $\xi_1, \xi_2, \ldots$ of "letters" from the "alphabet" $X=\{0,1,\ldots,p-1\}$. The variables $\xi_i$ may be considered as random variables or may not. By the random scramling of the finite string $\xi^n = \{\xi_1,\ldots,\xi_n\}$ we mean its modulo-$p$-product with a randomly choosen modulo-$p$ invertible $n \times n$ matrix with entries from $X$ .

To be more formal let us define $A_n$ as the set of all modulo-$p$-invertible $n \times n$ matrices with entries $a_{ij} \in X$ , $i,j=1,\ldots,n$. Then, choosing a matrix $A \in A_n$ with uniform distribution we get a random scrambling of $\xi = \{\xi_1,\ldots,\xi_n\}$ by performing the multiplication $A.\xi$ in modulo-$p$ arithmetics. We will generally suppose that $p$ is a prime number. In this case the operations in modulo-$p$-arithmetics are the operations on the residual field modulo $p$, and the invertible matrices are understood as the group of linear one to one transformations on that fields.

Suppose, that the sender and the reveiver agrees on a randomly choosen matrix $A \in \mathcal{A}_n$ i.e. they choose an element $A$ of $\mathcal{A}_n$ with uniform distribution. The encoding is carried out by breaking the message into blocks $\xi$ of length $n$, and then multiplying them by the matrix $A$ (in modulo arithmetics) in order to get the code-text $\underline{\eta}=A\xi$. The code text is sent over the channel and the receiver receives a possibly erroneus text $\tilde{\eta}$. It is supposed, however, that no sign is lost during the transmission. The receiver applies the inverse transformation $A^{-1}$ to the blocks of the received text to get the decoded text $\tilde{\xi}$. Then he is to decide if $\xi=\tilde{\xi}$ or not for any given block.

The expression "random scranbling" is motivated by the following lemma and Theorem 2.

<u>Lemma 1.</u> For all strings $\xi_1,\ldots,\xi_n$ and for all $x_1,\ldots,x_n$ the probability

$$P\{\eta_{i_1}=x_1,\ldots,\eta_{i_n}=x_n|\xi_1,\ldots,\xi_n\}$$

is independent of the permutation of the indeces $i_1,\ldots,i_n$.

<u>Proof:</u> If suffices to note that for all permutation-matrix $\Pi$ and for all invertible matrix $A$ the product $\Pi A$ is also an invertible matrix.

Let us denote the transmission error vector by $\underline{\varepsilon}$, i.e.

$$\varepsilon_i=\tilde{\eta}_i - \eta_i , \quad i=1,\ldots,n,$$

and suppose, that $p$ is prime number.

Then,

$$(1) \qquad \tilde{\xi} = A^{-1}\tilde{\eta} = A^{-1}(\eta+\varepsilon) = \xi+A^{-1}\varepsilon$$

From this and from Lemma 1 it follows that the probability

$$P\{\tilde{\xi}_{i_1}=x_1,\ldots,\tilde{\xi}_{i_n}=x_n \mid \underline{\xi},\ \varepsilon \neq 0\}$$

is also independent of the permutation of the indeces. Furthermore it can be shown that these conditional probabilities are the same for all $\varepsilon \neq 0$. Due to (1) it suffices to establish this fact for the conditional distribution of $A^{-1}\underline{\varepsilon}$, or, what is the same, for $A\,\underline{\varepsilon}$. We formulate this in the following theorem:

<u>Theorem 1:</u> If the random matrix $A$ takes its values in $\mathrm{A}_n$ with equal probabilities, then for all $\underline{\varepsilon}$, $\underline{\varepsilon} \neq 0$, the vector $A\underline{\varepsilon}$ is uniformly distributed over all non-zero (mod $p$) vectors of length $n$. (Since $A$ is invertible, $A\,\underline{\varepsilon}$ can not be zero.)

<u>Proof:</u> Let $\underline{\varepsilon}$, $\eta_o$ and $\eta_1$ be any fixed non-zero vectors, and consider the disjoint subsets $\mathrm{A}^o$, $\mathrm{A}^1$, $\subset \mathrm{A}_n$ with $A\underline{\varepsilon}=\eta_i$ iff $A\in\mathrm{A}^i$, $i=0,1$. Due to the uniformity assumption it suffices to prove that $\mathrm{A}^o$ and $\mathrm{A}^1$ has the same cardinality. For this reason let us fix a regular matrix $B$ with $B\eta_o=\eta_1$. Here neither $\eta_o$ nor $\eta_1$ is the zero-vector, therefore such a matrix $B$ always exists. Since $A\in\mathrm{A}^o$ implies $(BA)\,\underline{\varepsilon} = \eta_1$, i.e. $BA\in\mathrm{A}^1$, and since $BA_1 \neq BA_2$ if $A_1 \neq A_2$, we have got that the cardinality of $\mathrm{A}^1$ is not less than that of $\mathrm{A}^o$. Interchanging the role of $\eta_o$ and $\eta_1$ yields equality. The proof is completed.

Remark: If $p$ is not prime, then together with the identically zero vector we have to exclude any vector which has at least one component that is not co-prime to $p$. This is necessary condition for $A$ to be invertible. It is to see that the $A\underline{\varepsilon}$ $\underline{\varepsilon} \neq 0$, is uniformly distributed over the remaining set of vectors.

Corollary: Suppose $A\xi$ was sent and $\eta = A\xi + \varepsilon$ was received, where $\varepsilon \neq 0$, i.e. the transmission was in error. Then the decoded vector $\tilde{\xi} = A^{-1}\eta$ has the probability distribution

$$P(\tilde{\xi} = \tilde{x} \mid \xi = x, \; \underline{\varepsilon}) = \frac{1}{p^n - 1}, \quad \text{if} \quad \tilde{x} \neq x,$$

independently of $\underline{\varepsilon}$.

This corollary implies that the distribution of erroneus decoded texts is independent of the type of error.

Theorem 2. Let $P(\xi = x)$ be the distribution of the source block $\xi$ and suppose that the error vector $\underline{\varepsilon}$ is statistically independent of $\xi$. Then the distribution of the decoded block $\tilde{\xi}$ is given by

$$P(\tilde{\xi} = \tilde{x}) = \begin{cases} P(\xi = \tilde{x}) & \text{if} \quad \varepsilon = 0 \quad \text{(no error-case)} \\[2ex] \dfrac{1 - P(\xi = \tilde{x})}{p^n - 1} & \text{if} \quad \varepsilon \neq 0 \quad \text{(erroneus-case)} \end{cases}$$

Proof: The error-free case is obvious. Suppose, that $\underline{\varepsilon} \neq 0$.

Then

$$P(\tilde{\xi} = \tilde{x} \mid \varepsilon \neq 0) = \sum_{x} P(\tilde{\xi} = \tilde{x} \mid \xi = x, \; \varepsilon \neq 0) \, . \, P(\xi = x) =$$

$$= \frac{1}{p^n - 1} \sum_{x \neq \tilde{x}} P(\xi = x) = \frac{1 - P(\xi = \tilde{x})}{p^n - 1}$$

<u>Remark:</u> If $\xi$ and $\varepsilon$ are statistically dependent, then the distribution of $\tilde{\xi}$ may depend on the non-zero values of $\varepsilon$, as well: For $\varepsilon \neq 0$

$$P(\tilde{\xi}=\tilde{x}\,|\,\varepsilon)=\frac{1}{p^n-1}\cdot\sum_{x\neq\tilde{x}}P(\xi=x\,|\,\varepsilon)=\frac{1-P(\xi=\tilde{x}\,|\,\varepsilon)}{p^n-1}\quad.$$

<u>The hypothesis-testing problem:</u>

For the sake of simplicity, we will suppose that the error is independent of the message sent, i.e. $\xi_n$ and $\varepsilon_n$ are statistically independent. As we already said, we deal with the case when the receiver knows the distribution $P(\xi_n=x_n)$ of $\xi_n$. Decoding $\tilde{\xi}_n$, he is to decide if the transmission was error-free or not, i.e. $\tilde{\xi}_n=\xi_n$ or not. Since the matrix $A$ and, therefore, its inverse $A^{-1}$ was choosen randomly, he knows $\tilde{\xi}_n$'s distribution under both the hypothesis $H_o$ of error-free transmission and the hypothesis $H_1$ of erroneus transmission:

$$P(\tilde{\xi}_n=\tilde{x}_n\,|\,H_o) = P(\xi_n=\tilde{x}_n)$$

$$P(\tilde{\xi}_n=\tilde{x}_n\,|\,H_1) = \frac{1-P(\xi_n=\tilde{x}_n)}{p^n-1}\quad,$$

where $\tilde{x}_n$ is any $n$ -tuple with elements from $X=\{0,1,\ldots,p-1\}$, and $p$ is prime.

The solution of this simple vs. simple hypothesis testing problem is well known if the probabilities $P(\xi_n=x_n)$ are really known. Unfortunatelly, this is not the case if $\xi_n$ is the block of $n$ consequtive letters of any living language and, in particula those of written Hungarian. In this case we can use an approximation to $P(\tilde{\xi}_n=\tilde{x}_n\,|\,H_1)$ instead of its exact values. This is motivated by the fast convergence of $P(\xi_n=x_n)$ to zero with growing $n$, e.g. for $n=2$

$$\max_{x_2} P(\xi_2=x_2)=P(\xi_2=\text{Space}, A) \approx 0.025$$

(For statistics concerning written Hungarian, we refer to [4]).
For $n$ and $p$ large enough, the approximation

$$\frac{1}{p^n-1} \approx \frac{1}{p^n}$$

is sharp, too. Since in our case $p=31$, $n \gg 10$, we are going to replace

(2) $$P(\tilde{\xi}_n=\tilde{x}_n|H_1) \text{ by } P(\tilde{\xi}_n=\tilde{x}_n|\tilde{H}_1) = \frac{1}{p^n},$$

i.e. the distribution of the erroneus decoded text is supposed to be uniform.

In lack of even approximate knowledge of the statistical distribution of long blocks of written texts, we can make use of certain structures of it, which can be more easily investigated. Such a fine structure in written Hungarian is provided by the patterns of vowels and consonants.
Table 1. below gives the statistics of the number of runs from vowels in blocks of consecutive 40 letters in written Hungarian texts.
The space between words is also included in the alphabet, which, accordingly, consists of 31 letters including 9 vowels. (No run of length 4 or more was observed).

| No. of vowels (n) | Total | No.of Runs: R=n-i | | | |
|---|---|---|---|---|---|
| | | i=0 | i=1 | i=2 | i=3 |
| 10 | 9 | 9 | - | - | - |
| 11 | 31 | 29 | 2 | - | - |
| 12 | 73 | 53 | 18 | 2 | - |
| 13 | 154 | 113 | 37 | 4 | - |
| 14 | 266 | 171 | 85 | 9 | 1 |

| No. of vowels (n) | Total | No. of Runs: R=n-i | | | |
|---|---|---|---|---|---|
| | | i=0 | i=1 | i=2 | i=3 |
| 15 | 182 | 148 | 27 | 7 | - |
| 16 | 39 | 19 | 11 | 8 | 1 |
| 17 | 21 | 3 | 8 | 8 | 2 |
| 18 | 3 | - | 1 | 1 | 1 |
| Σ | 778 | 545 | 189 | 39 | 5 |
| % | | 0,701 | 0,243 | 0,050 | 0,006 |

Table 1: Distribution of the number of vowel runs.

Suppose now, that the decoded text contains $n$ vowels, and there was an error in the transmission. Under the modified hypothesis $\tilde{H}_1$ all vowel-not-vowel patterns are equally likely. For this case the distribution of the number of vowel-runs is known, see Mood [3]:

$$P(R=r) = \frac{\binom{n-1}{r-1}\binom{40-n+1}{r}}{\binom{40}{n}} \tag{3}$$

From this, it follows:

$$E(n-R) = \frac{n(n-1)}{40}$$

Since the expected number of the vowels

$$E(n) = 40 \cdot \frac{9 . 31^{39}}{31^{40}-1} \approx 11.61$$

There is a reason to expect $n-r$ larger than 1.

This gives rise to the following testing procedure:

Accept the hypothesis of erroneus transmission, if

- the number of vowels $\begin{cases} \text{is less than 10} \\ \text{is larger than 18} \end{cases}$
- there is a vowels-run of length 4 or more

- $10 \le n \le 12$ and $n-R \ge 2$
- $13 \le n \le 15$ and $n-R \ge 3$
- $16 \le n \le 18$ and $n-R \ge 4$

and detect no error otherwise.

The probability of an undetected error is given in Table 2, as a function of the number of vowels:

| n\i | 0 | 1 | 2 | 3 |
|---|---|---|---|---|
| 10 | 0,052 | 0,214 | 0,335 | 0,293 |
| 11 | 0,024 | 0,130 | 0,278 | 0,304 |
| 12 | 0,0092 | 0,068 | 0,197 | 0,296 |
| 13 | $3{,}1.10^{-3}$ | 0,030 | 0,108 | 0,220 |
| 14 | $8{,}6.10^{-4}$ | 0,011 | 0,058 | 0,161 |
| 15 | $1{,}9.10^{-4}$ | $3{,}4.10^{-3}$ | 0,023 | 0,087 |
| 16 | $3{,}2.10^{-5}$ | $7{,}8.10^{-4}$ | $7{,}4.10^{-3}$ | 0,038 |
| 17 | $3{,}9.10^{-6}$ | $1{,}3.10^{-4}$ | $1{,}7.10^{-3}$ | 0,012 |
| 18 | $2{,}9.10^{-7}$ | $1{,}5.10^{-5}$ | $2{,}9.10^{-4}$ | $2{,}9.10^{-3}$ |

Table 2: Table of some values of

$P(R=n-i), \quad i=0,1,2,3$

<u>Experimental results.</u> Table 1,2 can be used to determine the error-probabilities of both kinds of the test procedure suggested. We had considered, however, (more worthwile) to check our test on real data. Therefore we had choosen 250 consecutive blocks of length 40 of continuous written Hungarian. There were no blocks which were detected as erroneus.

In order to estimate the ratio of the srambled blocks which are considered error-free by the test, we needed random matrices. The generation of such matrices of size 40x40 is rather complicated, therefore we had decided to use "pseudo-random matrices". The idea of the generalization was motivated by the random-like statistical properties of linear Shift Register Sequences (LSRS). For a detailed treatment of the LSRS the reader is referred to the book by Golomb [1] . We recall that a LSRS over $GF(p)$ is a recurrence sequence, in which the elements are defined by a modulo $-p$-recursion:

$$(4) \qquad \Delta_{t+n} = \sum_{i=o}^{n-1} b_i \cdot \Delta_{t+i}\,, \quad b_o \neq o$$

with initials $\Delta_1,\ldots,\Delta_n$ and given constants $b_o,\ldots,b_{n-1}$ . The vectors $(\Delta_{t+1},\ldots,\Delta_{t+n})$ are the states of the Shift Register Generator (=LSRG). It is well-known that any LSRG partitiones the $n$ -length vectors over $GF(p)$ into cycles. If there is a cycle of length $p^n-1$ , then the LSRG is called of maximum length. The following lemma on LSRG 's helps to find a set of $n$ linearly independent vectors over $GF(p)$ .

<u>LEMMA 2:</u> Let $\Delta^o=(\Delta_1,\ldots,\Delta_n)$ be a vector in which all elements but $\Delta_n$ are zero, and consider a LSRG with $n$ -register. Then every $n$ consecutive states of the cycle to which $\Delta^o$ belongs are linearly independent vectors.

<u>Proof:</u> It is clear from the definition of the LSRS that the statement is valid for the first $n$ states. Let now $\Delta^i$ and $\Delta^{i+1}$ two consecutive states. It is clear, that there is a linear transformation $T$ with a non-degenerate matrix such that $\Delta^{i+1}=T\Delta^i$ , and, similarly, for any two states $\Delta$ and $\Delta^*$ of the same cycle there is an integer $j$ with $\Delta^*=T^j\Delta$ .

Therefore if $\Delta^i, \Delta^{i+1},\ldots,\Delta^{i+n-1}$ are $n$ consequtive states, then

$$\Delta^{i+t} = T^i\Delta^t, \quad t=o,1,\ldots,n-1,$$

i.e. $\Delta^i,\Delta^{i+1},\ldots,\Delta^{i+n-1}$ and $\Delta^o,\Delta^1,\ldots,\Delta^{n-1}$ are linearly independent at the same time.

<u>Corollary:</u> Any $n$ consecutive states of a LSRG of maximum lenght are linearly independent.

In the statistical investigation of the error-detecting capacity of the proposed method we had defined 10 pseudo-random matrices in the following way:

a) the coefficients $b_i$ in (4) were choosen randomly, and the 101st, 102nd,...,140th states of the corresponding LSRG following $\Delta^o$ were computed. This vectors yielded the rows. resp. the columns of a matrix $B$ resp. $B^*$ . According to Lemma 2, both $B$ and $B^*$ are invertible.

b) Non-zero numerals $(c_1,\ldots,c_n)$, $(d_1,\ldots,d_n)$ where choosen randomly, and a new matrix $A^{-1}$ was formed by defining the entries $a_{ij}$ of $A^{-1}$ by

$$a_{ij}=b_{ij} \cdot c_i \cdot d_j ,$$

where $b_{ij}$ are the entries of $B$ . Similarly, there was determined another matrix $A^{*-1}$ .

c) Steps a) and b) was repeated 5 times to get the 10 matrices. Obviously, there is no loss in considering the inverse matrices and thereby sparing computer time.

We have also choosen 10 error-patterns, randomly, more exactly 4 with 1 error, 3 with 2 errors, 2 with 3 errors and 1 with 4 errors. The error-vectors were multiplied by the generated inverse matrices and the resulting 100 vectors were stored. These vectors were added to all of the 250 text-blocks cyclically and the run-test was performed. Table 3 shows how frequently our test-procedure accepted the scrabled text to be written Hungarian. It can be seen from this table,

| Matrices | Error-patterns (No.of errors) | | | | | | | | | | $\Sigma$ |
|---|---|---|---|---|---|---|---|---|---|---|---|
| | 1 | | | | 2 | | | 3 | | 4 | |
| | 1 | 2 | 3 | 4 | 5 | 6 | 7 | 8 | 9 | 10 | |
| $A_1$ | 2 | 0 | 3 | 3 | 1 | 2 | 0 | 1 | 0 | 1 | 13 |
| $A_1^*$ | 2 | 3 | 2 | 1 | 2 | 2 | 3 | 0 | 1 | 0 | 16 |
| $A_2$ | 3 | 1 | 1 | 4 | 1 | 1 | 2 | 3 | 2 | 1 | 19 |
| $A_2^*$ | 0 | 1 | 0 | 2 | 2 | 3 | 2 | 0 | 1 | 2 | 13 |
| $A_3$ | 3 | 0 | 0 | 1 | 3 | 1 | 0 | 2 | 1 | 3 | 14 |
| $A_3^*$ | 0 | 3 | 2 | 3 | 4 | 1 | 4 | 3 | 2 | 0 | 22 |
| $A_4$ | 0 | 1 | 1 | 2 | 2 | 1 | 2 | 2 | 1 | 0 | 12 |
| $A_4^*$ | 3 | 3 | 2 | 2 | 3 | 3 | 0 | 2 | 0 | 1 | 19 |
| $A_5$ | 0 | 0 | 1 | 0 | 1 | 1 | 0 | 2 | 3 | 0 | 8 |
| $A_5^*$ | 0 | 3 | 0 | 0 | 3 | 3 | 2 | 1 | 0 | 1 | 13 |
| $\Sigma$ | 13 | 15 | 12 | 18 | 22 | 18 | 15 | 15 | 11 | 9 | 149 |

Table 3: Number of wrong decisions in the case of errors

that the relative frequency of wrong decisions in the case of transmission error is about 0,6%, an unexpectedly good result.

Concluding remarks:

The proposed method ensures reliable error-detection in transmission of written Hungarian. The coding-decoding algorithm is easily implementable, and the run-type test allows on-line error detection. The only weak point is that the method operates on relatively long blocks. Obviously, much shorter blocks could also ensure a reliable error-detection, but, in this case, more sophisticated test-procedures were needed, which, being very time--consuming, would prevent an on-line detection. Such a method could be,e.g. a dictionary-look-up, which supposes a large memory, as well. We should like to note, that even a theoretical limit on the block-length permitting a given level error detection is unknown.

Our experiments show that using purely run-like test, the block-length can not be essentially decreased, while keeping the on-line error-detection capacity acceptable. It is possible, however, to combine it with other simple test, e.g. with checking the frequencies of the 10 least frequent letters. Such combinations may result in a considerable decrease of the block length. Experiments in this respect are still in progress.

REFERENCES

[1] Golomb,S.W.: Shift register sequences, Holden-Day, San Francisco 1967.

[2] Hellman,M.E.: On using natural redundancy for error detection, IEEE Trans. on Communications, COM-22(1974) pp. 1690-1693.

[3] Mood,A.M.: The distribution theory of runs, Annals of Math. Stat. 11 (1940)pp.367-392.

[4] Nemetz,T., Szillèry,A.: Hungarian language-statistics, (in Hungarian), Alkalmazott Matematikai Lapok, (in print)

# RECURSIVE ESTIMATION IN THE "ALMOST SMOOTH" CASE

G. Pflug

Universität Wien

1. Introduction. Let $(\mathfrak{X}, \mathfrak{A})$ be a measurable space and $\{P_\theta\}_{\theta\varepsilon\Theta}$ a family of probability measures on $\mathfrak{A}$. We assume that the parameter space $\Theta$ is an open subset of the Euklidean space $\mathbb{R}^k$. Let $\{X_n\}_{n\varepsilon N}$ be a sequence of independent -valued random variables with distribution $P_{\theta_0}$ defined on some probability space $(\Omega, \mathcal{L}, \mu)$. The parameter $\theta_0$ is unknown and is to be estimated from the observations $\{X_n\}$.
A sequence $\{T_n\}_{n\varepsilon N}$ of estimates is called *recursive* if there are functions $\psi(n,T,x)$ defined on $(\mathbb{N} \times \Theta \times \mathfrak{X})$ such that

$$T_{n+1} = \psi(n, T_n, X_{n+1}) \qquad n > 0$$

A starting value $T_0$ is needed for this recursion, but it must not have any influence on the asymptotic behavior of $\{T_n\}$.

Several authors (Sakrison [6], Hasminskij [3], Fabian [2]) have considered recursive estimation procedures under various regularity conditions for the family $\{P_\theta\}$. The set of conditions is usually divided into two groups: Those, which ensure the consistency of $T_n$ (global conditions) and those, which give the asymptotic optimality (local conditions). The usual local regularity conditions are exactly the classical Cramér-Wald conditions, whereas the global conditions are rather specific and innatural. Therefore Hasminskij [3] and in the sequel Fabian [2] have defined procedures, which use auxiliary estimates for to get rid of the global conditions.

The organization of this paper is as follows: In section 2 we show that the assumption of the quadratic mean differentiability of the square roots of the densities can replace all other local conditions. In section 3 we consider the "almost smooth" case introduced by Ibragimov/Hasminskij [4]. In this case, the $L^2$-differentiability assumption is violated, but nevertheless there is a "quick" convergent recursive estimation procedure available.

## 2. The "smooth" case.

Let us state the following set of assumptions:

(2.1) Assumption.

(i) There is a $\sigma$-finite measure $\nu$ dominating the family $\{P_\theta\}$. Let

$$f_\theta(x) = f(\theta,x) := \frac{dP_\theta}{d\nu}(x)$$

(ii) The mapping $\theta \to \sqrt{f_\theta}$ is differentiable in the $L^2(\nu)$ sense. We denote the derivative by $g(\theta,x)$ (Note that $g_\theta$ is a k-vector of $L^2(\nu)$-functions and $g'_\theta$ is the transposition of $g_\theta$). Let $I(\theta) = 4.\int g_\theta g'_\theta d\nu$ (componentwise integration is used).

(iii) Let

$$k(\theta,x) = \begin{cases} \dfrac{g(\theta,x)}{\sqrt{f(\theta,x)}} & f(\theta,x) > 0 \\ 0 & \text{otherwise} \end{cases}$$

There is a neighbourhood U of $\theta_0$ such that

$k_\theta \to L^2(P_{\theta_0})$ (componentwise) for $\theta \in U$

and

$k_\theta \to K_{\theta_0}$ in the sense of $L^2(P_{\theta_0})$ as $\theta \to \theta_0$.

(iv) $\theta \to I(\theta)$ is continuous at $\theta_0$ and $I(\theta_0)$ is nonsingular.

(2.2) Assumption. Let $S_n$ be an auxiliary estimate, such that

$$\gamma_n \, \|S_n - \Theta_0\|_{R^k} \to 0 \qquad P^{\mathbb{N}}_{\theta_0} \text{ - a.e.}$$

for a nonnegative sequence $\gamma_n \to \infty$.

If $C$ is a closed convex set in $\mathbb{R}^k$, the convex projection $Pr_C$ onto $C$ is well defined. If $C$ is a closed sphere with center $S_n$ and radius $\rho_n$, the projection operator onto $C$ is denoted be $Pr_{S_n,\rho_n}(\cdot)$.

<u>(2.3) The procedure.</u> The recursive estimate $T_n$ is defined as follows:

$$(1) \quad T_{n+1} = Pr_{S_n,\rho_n}[T_n + n^{-1}.I^{-1}(T_n)2k(T_n,X_{n+1})]$$

where $I^{-1}(T_n) = E_k$, the k x k identy matrix, if $\det I(T_n) = 0$.

<u>(2.4) Theorem.</u> Let the assumptions (2.1) and (2.2) be satisfied. If $\rho_n = \max(n^{-\gamma},\gamma_n^{-1})$ and $\gamma < 1/2$ then for the recursive estimate $T_n$ given by (1) it is true that

$$\mathcal{L}(n^{1/2}(T_n-\theta_0)|P^n_{\theta_0}) \to N(0,I^{-1}(\theta_0))$$

<u>Proof.</u> There are only slight changes of Fabian's proof([2], Theorem 3.2) necessary.
First of all we show that

$$(2) \quad \int k(\theta,x)f(\theta,x)\, d\nu(x) = \underline{0}$$

Let $\theta_1$ be an arbitrary element of $R^k$. The proof of (2) follows from the equations:

$$\theta_1'.\int k_\theta.f_\theta\, d\nu = \theta_1'.\int g_\theta.\sqrt{f_\theta}\, d\nu =$$

$$= \lim_{t\to 0} \frac{1}{t} \int(\sqrt{f_{\theta+t.\theta_1}} - \sqrt{f_\theta})\sqrt{f_\theta}\, d\nu =$$

$$= \lim_{t\to 0} \frac{1}{2t} \int(\sqrt{f_{\theta+t.\theta_1}} - \sqrt{f_\theta})^2 d\nu =$$

$$= \lim_{t\to 0} \frac{t}{2} . \| \theta_1'.g_\theta \|^2_{L^2(\nu)} = 0$$

Next let $M(\theta) = \int k(\theta,x)\, f(\theta_0,x)d\nu$.

Because of Assumption (2.1) (iii) $M(\theta)$ is well defined for $\theta \in U$. As we have just shown $M(\theta_0) = \underline{0}$. We now claim that for the derivative $D(M(.))(\theta_0)$ at the point $\theta_0$ we have

(3) $\quad D(M(.))(\theta_0) = -I(\theta_0) \,/\, 2$

For to prove (3) we note that for $\theta \in U$

$$M(\theta) = \int k_\theta (f_{\theta_0} - f_\theta) d\nu = \int k_\theta (f_{\theta_0} - \sqrt{f_{\theta_0} f_\theta} + \sqrt{f_{\theta_0} f_\theta} - f_\theta) d\nu =$$

$$= \int k_\theta . \sqrt{f_{\theta_0}} (\sqrt{f_{\theta_0}} - \sqrt{f_\theta}) d\nu + \int g_\theta (\sqrt{f_{\theta_0}} - \sqrt{f_\theta}) d\nu$$

and hence by Assumption (2.1) (iii)

$$\lim_{\theta \to \theta_0} \| \theta - \theta_0 \|^{-1}_{R_k} . \| M(\theta) + 2 \int g_{\theta_0} g'_{\theta_0} d\nu . (\theta - \theta_0) \|_{R_k} = 0$$

whence (3) follows.

Furthermore according to Assumption (2.1) (iii) $k(T_n, .) \in L^2(P_{\theta_0})$ eventually and $\| k(T_n, .) \|_{L^2(P_{\theta_0})} \rightarrow I(\theta_0)$.

The notion "eventually" is meant in the sense of Fabian [2].

Hence the assertion of the theorem follows from Fabian's Theorem 3.2 in [2].

## 3. The "almost smooth" case

In [4], Ibragimov and Has'minskij have found out the remarkable fact, that there are examples of parametric families $\{P_\theta\}$ such that the maximum likelihood estimator $\hat{\theta}_n$ satisfies

(4) $\quad \mathcal{L}(\sqrt{n \log n}\, (\hat{\theta}_n - \theta_0) | P^n_{\theta_0}) \rightarrow N(0, 2/B)$

The authors called those cases "almost smooth", since besides the difference in the norming sequence $\sqrt{n \log n}$ the estimators behave like in the smooth case.

Let us assume in the following for the sake of simplicity, that the unknown parameter is one dimensional.

It is common to all examples for (4) that the space $\mathfrak{X}$ is decomposable into two parts $\mathfrak{X} = \mathfrak{X}^{(1)} \cup \mathfrak{X}^{(2)}$ such that $\nu(\mathfrak{X}^{(1)}) < \infty$ and $\theta \longmapsto \sqrt{f_\theta}\big|_{\mathfrak{X}^{(2)}}$ is $L^2$-differentiable at the point $\theta_0$ whereas

$\theta \longmapsto \sqrt{f_\theta}\big|_{\mathfrak{X}^{(1)}}$ is only $L^p$-differentiable for all $p < 2$ at $\theta_o$. For asymptotic considerations we can assume w.l.o.g. that $\mathfrak{X}^{(2)} = \emptyset$. Besides that we need a couple of regularity conditions:

(3.1) Assumptions.

(i) There is a finite measure $\nu$ dominating the family $\{P_\theta\}$. Let $f_\theta(x) := f(\theta,x) := \frac{dP_\theta}{d\nu}(x)$. We assume further that the family $\{f_\theta\}$ is uniformly bounded in $L^\infty(\nu)$.

(ii) The mapping $\theta \longmapsto \sqrt{f_\theta}$ is differentiable in the $L^1$-sense. We denote the derivative by $g(\theta,x)$.

(iii) We shall use the following notation:

$$g(t,\theta,x) := g(\theta,x).1_{\{|g(\theta,x)| \leqslant t\}} \qquad t > 0$$

and assume that there is a constant $C_1$ such that for sufficiently large $t$

$$\left\| \frac{g(t,\theta,\cdot)}{\sqrt{f(\theta,\cdot)}} \right\|_{L^\infty(\nu)} \leqslant C_1.t$$

(iv) Let there exist another constant $C_2$ such that

$$\left\| \frac{\sqrt{f_\theta}-\sqrt{f_{\theta_o}}}{\theta-\theta_o} - g(|\theta-\theta_o|^{-1/2},\theta_o,\cdot)\right\|_{L^2(\nu)} \leqslant C_2$$

for all $\theta$ in a neighborhood of $\theta_o$.

(v) We assume the existence of the following limit uniformly in a neighborhood of $\theta_o$:

$$\lim_{t\to\infty} \frac{\|g(t,\theta,\cdot)\|^2_{L^2(\nu)}}{\log t} =: I(\theta)$$

Let further $\theta \longmapsto I(\theta)$ be continuous and $I(\theta_o) > 0$.

(3.2) Examples.

(i) Let $\mathbb{X} = \mathbb{R}$ and $\nu = \lambda$ (Lebesgue measure).
The location parameter family $\{P_\theta\}$ given by the densities

$$f(\theta,x) := C.\exp(-|x-\theta|^{1/2}) \qquad \theta \in \mathbb{R}$$

satisfies (3.1) and $I(\theta) \equiv C/8$.

(ii) The same is true for the triangle distribution family:

$$f(\theta,x) := (\sqrt{a}-a|\theta-x|)^+ \qquad a > 0$$

In this case $I(\theta) \equiv a/2$.

(iii) Similar examples can be found by taking linear combinations of densities of the type (i) and (ii), the weighting functions being $L^2$-differentiable (see [3]).

(3.3) Remark. Let $d(\theta,\theta_0)$ denote the Hellinger distance, i.e.

$$d^2(\theta,\theta_0) = \frac{1}{2} \int (\sqrt{f_\theta}-\sqrt{f_{\theta_0}})^2 d\nu.$$

If assumption (3.1) is satisfied then

$$\lim_{\theta\to\theta_0} \frac{d_2{}^2(\theta,\theta_0)}{(\theta-\theta_0)^2(-\log|\theta-\theta_0|)} = I(\theta_0)/4$$

This is an easy consequence of the assumptions (3.1) (iv) and (v).

(3.4) Assumption. Let $S_n$ be an auxiliary estimate, such that

$$n^\gamma|S_n - \theta_0| \to 0 \qquad P_{\theta_0}^{\mathbb{N}} \text{ - a.e.}$$

for a $\gamma > 0$.

(3.5) The procedure.

Let $\varepsilon := \min(\gamma,1/2)$
and

$$k(n,\theta,x) := \frac{g(n^{\varepsilon/2},\theta,x)}{\sqrt{f(\theta,x)}} - c_n(\theta)$$

where

$$c_n(\theta) := \int g(n^{\varepsilon/2},\theta,x)\sqrt{f(\theta,x)}\,d\nu(x)$$

The recursive estimate $\{T_n\}$ is defined by

$$(5)\qquad T_{n+1} = Pr_{S_n,\rho_n}[T_n+(\varepsilon.n.\log n)^{-1}I^{-1}(T_n).k(n,T_n,X_{n+1})]$$

where $\rho_n = \frac{1}{2}n^{-\varepsilon}$. $I(\theta)$ can be defined arbitrarily $\neq 0$ for those $\theta$, for which the limit (3.1) (v) does not exist.

(3.6) Lemma. Let the assumptions (3.1) and (3.4) be satisfied. Then

$$\sqrt{n}(T_n - \theta_o) \to 0 \qquad P_{\theta_o}^N \text{ - a.e.}$$

Proof. Let $A_m = \{\omega \mid |S_k(\omega) - \theta_o| < \frac{1}{2}k^{-\varepsilon} \quad k > m\}$
According to (3.4) $P(A_m) \to 1$ as $m \to \infty$. It suffices to consider the restrictions to the sets $A_m$. On $A_m$ we have for $n > m$:

$$(6)\qquad (T_{n+1}-\theta_o)^2 \leqslant (T_n-\theta_o)^2 + \frac{2(T_n-\theta_o)}{n.\varepsilon.I(T_n)}\;\frac{k(n,T_n,X_{n+1})}{\log n} +$$

$$+ \frac{1}{n^2.\varepsilon^2.I^2(T_n)}\;\frac{k^2(n,T_n,X_{n+1})}{\log^2 n}$$

Now let

$$M(n,\theta) := E(k(n,\theta,\cdot)|P_{\theta_o})$$

Then $M(n,\theta_0) = 0 \ \forall n$ according to the definition. We show further, that if $|\theta-\theta_0| < n^{-\varepsilon}$:

$$(7)\begin{cases} M(n,\theta) = \alpha_n(\theta).(\theta-\theta_0) & \text{where} \\ \alpha_n = -\varepsilon.I(\theta).\log n + o(\log n) \\ \text{and } o \text{ does not depend on } \theta. \end{cases}$$

For to prove (7) notice that

$$M(n,\theta) = \int \left[\frac{g(n^{\varepsilon/2},\theta,x)}{\sqrt{f(\theta,x)}} - c_n(\theta)\right] f(\theta_0,x)d\nu(x) =$$

$$= \int \frac{g(n^{\varepsilon/2},\theta,x)}{\sqrt{f(\theta,x)}} (f(\theta_0,x)-f(\theta,x))d\nu(x) =$$

$$= 2(\theta-\theta_0)\int g(n^{\varepsilon/2},\theta,\cdot)\left(\frac{\sqrt{f_{\theta_0}}-\sqrt{f_\theta}}{\theta-\theta_0}\right)d\nu +$$

$$+ \int \frac{g(n^{\varepsilon/2},\theta,\cdot)}{\sqrt{f_\theta}} (\sqrt{f_{\theta_0}} - \sqrt{f_\theta})^2 d\nu =$$

$$= M_1(n,\theta) + M_2(n,\theta) + M_3(n,\theta)$$

where

$$M_1(n,\theta) = -2(\theta-\theta_0)\int g(n^{\varepsilon/2},\theta,\cdot)g(|\theta-\theta_0|^{-1/2},\theta,\cdot)d\nu$$

$$M_2(n,\theta) = 2(\theta-\theta_0)\int g(n^{\varepsilon/2},\theta,\cdot)\left[g(|\theta-\theta_0|^{-1/2},\theta,\cdot)-\frac{\sqrt{f_\theta}-\sqrt{f_{\theta_0}}}{\theta-\theta_0}\right]d\nu$$

$$M_3(n,\theta) = \int \frac{g(n^{\varepsilon/2},\theta,\cdot)}{\sqrt{f_\theta}} (\sqrt{f_{\theta_0}}-\sqrt{f_\theta})^2 d\nu$$

Under the assumption $|\theta-\theta_0| < n^{-\varepsilon}$ it follows that

$$M_1(n,\theta) = -2(\theta-\theta_o) \int g^2(n^{\varepsilon/2},\theta,\cdot)d\nu =$$

$$= -2(\theta-\theta_o)\ I(\theta).(\log(n^{\varepsilon/2}) + o(\log n)) =$$

$$= -\varepsilon.(\theta-\theta_o)[I(\theta).\log n + o(\log n)]$$

Because of assumption (3.1) (iv) and Schwarz's inequality we get

$$M_2(n,\theta) = (\theta-\theta_o).O(\sqrt{\log n})$$

The third expression $M_3$ can be treated using assumption (3.1) (iii) and remark (3.3)

$$|M_3(n,\theta)| \leqslant C_1.n^{\varepsilon/2}.\|\sqrt{f_{\theta_o}}-\sqrt{f_\theta}\|^2_{L^2(\nu)} =$$

$$= O(n^{\varepsilon/2}).O(n^{-\varepsilon}.\log n).|\theta-\theta_o| = o(|\theta-\theta_o|)$$

We have therefore established (7).
We further show a little bit more than we actually need, namely that in a neighborhood of $\theta_o$

$$(8)\qquad E(k^2(n,\theta,\cdot)|P_{\theta_o}) = \varepsilon/2.I(\theta).\log n + O(1)$$

where O does not depend on $\theta$.
The argumentation goes as follows

$$E(k^2(n,\theta,\cdot)|P_{\theta_o}) = \int\left[\frac{g(n^{\varepsilon/2},\theta,\cdot)}{\sqrt{f_\theta}} - c_n(\theta)\right]^2 f(\theta_o,\cdot)d\nu =$$

$$= \int [\frac{g(n^{\varepsilon/2},\theta,\cdot)}{\sqrt{f_\theta}} - c_n(\theta)]^2.f(\theta,\cdot)d\nu +$$

$$+ \int [\frac{g(n^{\varepsilon/2},\theta,\cdot)}{\sqrt{f_\theta}} - c_n(\theta)]^2(\sqrt{f_{\theta_o}}+\sqrt{f_\theta})(\sqrt{f_{\theta_o}}-\sqrt{f_\theta})d\nu =$$

$$= E_1 + E_2$$

Since $c_n(\theta) = \int g(n^{\varepsilon/2},\theta,\cdot)\sqrt{f_\theta}d\nu \to \int\ g(\theta,\cdot)\sqrt{f_\theta}d\nu = 0$

as $n \to \infty$ and $\int g^2(n^{\varepsilon/2},\theta,\cdot)d\nu = \varepsilon/2\ I(\theta)\log n + o(\log n)$ it follows that

$$E_1 = \varepsilon/2\cdot I(\theta)\cdot\log n + o(\log n)$$

whereas (because of (3.1) (iii))

$$|E_2| < \left\| \frac{g(n^{\varepsilon/2},\theta,\cdot)}{\sqrt{f_\theta}} - c_n(\theta)\right\|^2_{L^\infty(\nu)} \cdot \|\sqrt{f_{\theta_0}}+\sqrt{f_\theta}\|_{L^\infty(\nu)} \|\sqrt{f_{\theta_0}}-\sqrt{f_\theta}\|_{L_1} =$$

$$= O(n^\varepsilon).O(1).O(n^{-\varepsilon}) = O(1)$$

Hence (8) is established.

Let now $\mathbb{B}_n$ denote the $\sigma$-algebra generated by $T_1,\ldots,T_n$. We take conditional expectations on both sides of (6), use (7) and (8) and get

$$E((T_{n+1}-\theta_0)^2|\mathbb{B}_n) = (T_n-\theta_0)^2(1-\tfrac{2}{n}+o(\tfrac{1}{n})) + O(\tfrac{1}{n^2})$$

Let $\beta_n := n(T_n - \theta_0)^2$. Then

$$E(\beta_{n+1}|\mathbb{B}_n) = \beta_n(1-\tfrac{1}{n}+o(\tfrac{1}{n})) + O(\tfrac{1}{n^2})$$

Well known martingale theorems now imply that $\beta_n$ converges a.e. and $\Sigma\frac{\beta_n}{n} < \infty$ a.e., whence $\beta_n \to 0$ a.e.

<u>(3.7) Theorem.</u>

Let the assumptions (3.1) and (3.4) be satisfied. Then

$$\mathcal{L}(\sqrt{n\log n}\,(T_n-\theta_0)|P^n_{\theta_0}) \to N(0,(2\varepsilon I(\theta_0))^{-1})$$

<u>Proof.</u> We adopt the notation of lemma (3.6). Whenever $|T_n-\theta_0| < \frac{1}{2}n^{-\varepsilon}$ and $|S_n-\theta_0| < \frac{1}{2}n^{-\varepsilon}$ then

$$(9) \qquad T_{n+1} = T_n + \frac{M(n,T_n)}{\varepsilon.n.\log n\ I(T_n)} + \frac{k(n,T_n,X_{n+1})-M(n,T_n)}{\varepsilon.n.\log n\ I(T_n)}$$

Because of lemma (3.6) for every $\omega$, there is a $N(\omega)$ such that (9) is true for all $n > N(\omega)$.
Inserting (7) into (9) we get

$$(T_{n+1}-\theta_0) = (T_n-\theta_0)(1-\tfrac{1}{n}+o(\tfrac{1}{n})) + \tfrac{1}{n}.Z_n$$

where $Z_n = \dfrac{k(n,T_n,X_{n+1})-M(n,T_n)}{\varepsilon.I(T_n).\log n}$

We have $E(Z_n|\mathcal{B}_n) = 0$
and because of (8):

$$(10) \quad E(Z_n^2|\mathcal{B}_n) = (2.\varepsilon.I(T_n).\log n)^{-1}+o(\log n)^{-1}$$

By standard argumentation, the limiting distribution of $\sqrt{n \log n}\ (T_n-\theta_0)$ coincides with that of

$$\sqrt{n \log n}\ .\ \sum_{j=2}^{n} \frac{Z_j}{j} \prod_{i=j+1}^{n} (1-\tfrac{1}{i}) = \sqrt{\frac{\log n}{n}} \sum_{j=2}^{n} Z_j$$

In order to make use of Dvoretzky's central limit theorem ([1], Theorem 2.2) we have to show that

$$(11) \quad \frac{\log n}{n} \sum_{j=2}^{n} E(Z_j^2 | \mathcal{B}_j) \xrightarrow{P} (2\varepsilon.I(\theta_0))^{-1}$$

and

$$(12) \quad \frac{\log n}{n} \sum_{j=2}^{n} E(Z_j^2.1_{\{|Z_j|>\sqrt{\frac{n}{\log n}}.\eta\}} | \mathcal{B}_j) \xrightarrow{P} 0$$

for every $\eta > 0$.

Since $\sum_{j=2}^{n} \dfrac{1}{\log j} = \dfrac{n}{\log n} + o(\dfrac{n}{\log n})$ and $I(T_n) \to I(\theta_0)$ a.s. (11) follows from (10).

For to show (12) notice that

$$\| Z_n \|_{L^\infty(\nu)} < C_1 . n^{\varepsilon/2} . O(\frac{1}{\log n}) + o(\frac{1}{\log n})$$

Since $\varepsilon<1/2<1$, there is for every $\eta > 0$ an $m \in N$ such that $\| Z_j \|_{L^\infty(\nu)} < \eta . \sqrt{\frac{n}{\log n}}$ for all $j < n$ and all $n > m$.

Hence (12) is shown and the desired result is established.

(3.8). Remark.

An optimal recursive estimator is found as follows: Take an arbitrary auxiliary estimate $S_n^{(1)}$ satisfying (3.4) with a $\gamma > 0$. Use this estimate to construct an recursive estimate $S_n^{(2)}$ satisfying $n^{1/2}(S_n^{(2)}-\theta_0) \to 0$ a.s. (Lemma 3.6). Use this estimate as an auxiliary estimate and use for a second time the procedure (3.6), now with $\varepsilon = 1/2$. The resulting estimate $T_n$ satisfies then

$$\mathcal{L}(\sqrt{n \log n}\ (T_n-\theta_0)|P^n_{\theta_0}) \to N(0,I^{-1}(\theta_0))$$

The asymptotic variance $I^{-1}(\theta_0)$ is exactly the same which appears in Ibragimov/Has'minskij [4]. (Their constant B equals our $2.I(\theta_0)$).

References


[1] Dvoretzky A. (1970). Asymptotic normality for sums of dependent random variables. Sixth. Berkeley Symposium.

[2] Fabian V. (1978). On asymptotically efficient recursive estimation. Ann. Statist. Vol.6, No.4, 854-866.

[3] Has'minskij R.Z. (1974). Sequential estimation and recursive asymptotically optimal procedures of estimation and observation control. Proc. Prague Symp. Asymptotic Statist. (Charles University Prague) 1, 157-178.

[4] Ibragimov I.A., Has'minskij R.Z. (1973). Asymptotic analysis of statistical estimators for the "almost smooth" case. Theor. Prob. Appl. Vol. 18, No.2, 241-252.

[5] LeCam L. (1970). On the assumptions used to prove asymptotic normality of maximum likelihood estimates. Ann. Math. Stat. Vol. 41,No. 3, 802-828.

[6] Sakrison D.J. (1966). Stochastic approximation: A recursive method for solving regression problems. Adv. Communication Systems 2, 51-106.

# HOW SMALL ARE THE INCREMENTS OF A WIENER SHEET?

by

P. RÉVÉSZ

## I. INTRODUCTION

In [1] we investigated the question: "How small are the increments of a Wiener process?" The main answer of this questions can be summarized in the following:

THEOREM A. (Csörgő-Révész, 1979) Let $\{W(t),\ t\geq 0\}$ be a Wiener process and consider the process

$$I(T)=\inf_{0\leq t\leq T-a_T}\ \sup_{0\leq s\leq a_T}|W(t+s)-W(t)| \qquad (T>0)$$

where $a_T$ is a non-decreasing function of $T$ for which

(i) $0<a_T\leq T$ ,

(ii) $T^{-1}a_T$ is non-decreasing.

Then

$$\liminf_{T\to\infty}\gamma_T I(T)=1 \qquad \text{a.s.}$$

where

$$\gamma_T=\left(\frac{8(\log Ta_T^{-1}+\log\log T)}{\pi^2 a_T}\right)^{1/2}$$

If we also have

$$\text{(iii)}\qquad \lim_{T\to\infty}\frac{\log Ta_T^{-1}}{\log\log T}=\infty$$

then

$$\lim_{T\to\infty} \gamma_T \, I(T) = 1 \qquad \text{a.s.}$$

Here we mention only two trivial consequences of this Theorem. The first one can be obtained from Theorem A by the choice $a_T = C\log T$ .

CONSEQUENCE A.1. For all $T$ big enough, for all $0<\delta<\varepsilon$ and for almost all $\omega\in\Omega$ (the basic space) there exists a $0\leq t = t(\varepsilon,\delta,\omega,T) \leq T-a_T$ such that

$$\sup_{0\leq s\leq a_T} |W(t+s) - W(t)| \leq \varepsilon+\delta$$

but for all $0\leq t\leq T-a'_T$

$$\sup_{0\leq s\leq a_T} |W(t+s) - W(t)| \geq \varepsilon-\delta \ .$$

where

$$a_T = \frac{8\varepsilon^2}{\pi^2} \log T \ .$$

An other form of this Consequence can be given by introducing the following

Definition. A function $f(x)$ is called $\varepsilon$ -constant on an interval $[a,b]$ if

$$\sup_{a\leq x\leq b} |f(x) - f(a)| \leq \varepsilon \quad .$$

Then we have

CONSEQUENCE A.1 *. For all $T$ big enough, for all $0<\delta<\varepsilon$ and for almost all sample function of $W(t)$ there exists a random interval $[t,t+a_T]$ of size $a_T = \frac{8\varepsilon^2}{\pi^2}\log T$ where $W(t)$ is $(\varepsilon+\delta)$ - constant.

Our second Consequence follows from Theorem A by the choice

$a_T = T$ .

CONSEQUENCE A.2. We have

$$\liminf_{T\to\infty} \left(\frac{8\log\log T}{\pi^2 T}\right)^{1/2} \sup_{0\le s\le T} |W(s)| = 1 \qquad \text{a.e.}$$

This Consequence was proved originally by Chung (1948).

In case of a Wiener sheet to find the analogue of Theorem A seems to be quite hard. In this paper we only intend to give the analgous of Consequences A.1 and A.2.

However these results will be not so exact as they are in the case of the one-parameter Wiener process.

## 2. AN INEQUALITY

The proof of Theorem A is based on the following well-known

THEOREM B. <u>We have</u>

$$P\{\sup_{0\le t\le 1} |W(t)| \le u\} =$$

$$(1) \qquad = \frac{1}{\sqrt{2\pi}} \int_{-u}^{+u} \sum_{k=-\infty}^{+\infty} (-1)^k \exp\left(-\frac{(x-2ku)^2}{2}\right) dx =$$

$$= \frac{4}{\pi} \sum_{k=0}^{\infty} \frac{(-1)^k}{2k+1} \exp\left(-\frac{\pi^2(2k+1)^2}{8u^2}\right) .$$

The first sum of (1) can be applied successfully if $u$ is big. For small $u$ the application of the second sum is more advisable. In fact we have

CONSEQUENCE B.1. <u>For any</u> $\varepsilon>0$ <u>there exists a</u> $u_0 = u_0(\varepsilon) > 0$ <u>such that</u>

$$\frac{1-\varepsilon}{u}\sqrt{\frac{2}{\pi}}\, e^{-\frac{u^2}{2}} \le P\{\sup_{0\le t\le 1}|W(t)| \ge u\} \le \frac{1}{u}\sqrt{\frac{2}{\pi}}\, e^{-\frac{u^2}{2}}$$

<u>if</u> $u\ge u_0$ .

CONSEQUENCE B.2. For any $u>0$ we have

$$\frac{4}{\pi}\left(e^{-\frac{\pi^2}{8u^2}} - \frac{1}{3} e^{-\frac{9\pi^2}{8u^2}}\right) \le P\{\sup_{0\le t\le 1}|W(t)| \le u\} \le \frac{4}{\pi} e^{-\frac{\pi^2}{8u^2}}$$

and if $u\le 4$ then we also have

$$e^{-\frac{\pi^2}{8u^2}} \le \frac{4}{\pi}\left(e^{-\frac{\pi^2}{8u^2}} - \frac{1}{3} e^{-\frac{9\pi^2}{8u^2}}\right) .$$

Let now $\{W(x,y);\ x\ge 0,\ y\ge 0\}$ be a Wiener sheet, i.e. a continuous Gaussian process with

$$EW(x,y) = 0$$

and

$$EW(x_1,y_1)W(x_2,y_2) = \min(x_1,x_2)\ \min(y_1,y_2) .$$

The exact distribution of

$$M = \sup_{\substack{0\le x\le 1\\ 0\le y\le 1}} |W(x,y)|$$

is unknown. An analogue of Consequence B.1 is known and it says:

THEOREM C. (Goodman, 1976) <u>For any</u> $\varepsilon>0$ <u>there exists a</u> $u_0=u_0(\varepsilon)$ <u>such that</u>

$$\frac{4-\varepsilon}{u}\sqrt{\frac{2}{\pi}}\, e^{-\frac{u^2}{2}} \le P\{M \ge u\} \le \frac{4}{u}\sqrt{\frac{2}{\pi}}\, e^{-\frac{u^2}{2}}$$

<u>if</u> $u\ge u_0$ .

In fact the proof of Theorem A is based on Consequence B.2 rather than Consequence B.1. At present our main difficulty is to find the analogue of Consequence B.2 for a Wiener sheet. Unfortunately we can give a very weak result.

THEOREM 1. There exist positive constants $C_1, C_2, C_3, C_4$ such that

$$C_1 \exp\{-C_2 u^{-2} (\log 1/u)^5\} \le P\{M \le u\} \le$$

$$\le C_3 \exp\{-C_4 u^{-2} \log 1/u\} .$$

REMARK 1. This result shows the intuitively clear fact that the distribution function of $\sup_x |W(x)|$ and that of $\sup_{x,y} |W(x,y)|$ are much more different near to $0$ than at the tail.

REMARK 2. The power 5 of $\log 1/u$ in Theorem 1 can be easily changed by a smaller one but I cannot get the exact one.

The proof of Theorem 1 is based on several lemmas. The first one is quite trivial.

LEMMA 1. Let $(X_1, X_2, \ldots, X_n)$ be a Gaussian random vector with $EX_i = 0 \quad (i=1,2,\ldots,n)$ . Then for any $u>0$, $u_1, u_2, \ldots, u_p$ reals and $p=1,2,\ldots,n-1$ we have

$$P\{|X_{p+1}| \le u, \quad |X_{p+2}| \le u, \ldots, |X_n| \le u \mid X_1 = X_2 = \ldots = X_p = 0\} \ge$$

$$\ge P\{|X_{p+1}| \le u, \quad |X_{p+2}| \le u, \ldots, |X_n| \le u \mid X_1 = u_1, X_2 = u_2, \ldots, X_p = u_p\}$$

with probability 1.

A HINT TO THE PROOF. The conditional covariance matrix of $X_{p+1}, X_{p+2}, \ldots, X_n$ given $X_1 = u_1,\ X_2 = u_2, \ldots, X_p = u_p$ does not depend on the actual values of $u_1, u_2, \ldots, u_p$ but the conditional expectation does.

The next Lemma is a simple consequence of the above one.

LEMMA 2. Let $u>0$ and let $f_1(y), f_2(y), \ldots, f_p(y)$ $(0\le y\le 1;\ p=1,2,\ldots,n-1)$ be a sequence of continuous functions. Then for any $0\le x_i\le 1$ $(i=1,2,\ldots,n)$ we have

$$P\{\| W(x_{p+1},\cdot)\| \le u,\ \| W(x_{p+2},\cdot)\| \le u,\ldots,\| W(x_n,\cdot)\| \le u \mid W(x_1,y)\equiv W(x_2,y)\equiv\ldots\equiv W(x_p,y)\equiv 0\} \ge$$

$$\ge P\{\| W(x_{p+1},\cdot)\| \le u,\ \| W(x_{p+2},\cdot)\| \le u,\ldots,\| W(x_n,\cdot)\| \le u \mid W(x_1,y)=f_1(y),\ W(x_2,y)=f_2(y),\ldots,W(x_p,y)=f_p(y)\}$$

with probability 1 where $W(x,y)$ is a Wiener sheet and $\| W(x,\cdot)\| = \sup_{0\le y\le 1} |W(x,y)|$.

Our next Lemma is also tirivial without any proof.

LEMMA 3. For any $0<x_0\le 1$ we have

$$P\{\| W(\tfrac{x_0}{2},\cdot)\| \le u \mid W(x_0,y)\equiv 0\} = P\{\frac{\sqrt{x_0}}{2} \| W(\cdot)\| \le u\}.$$

LEMMA 4. For any integer $n$ we have

$$P\{\max_{i\le k\le 2^n} \| W(\tfrac{k}{2n},\cdot)\| \le u\} \le P\{\| W(\cdot)\| \le u\} \prod_{k=0}^{n-1} (P\{\sqrt{\frac{1}{2^{k+2}}}\| W(\cdot)\| \le u\})^{2^k}$$

PROOF. Put

$$A_u = \{f(y)\in C(0,1): \| f(y)\| \le u\}$$

and consider

$$P\{\max_{1\le k\le 2^n} \| W(\tfrac{k}{2^n}, \cdot)\| \le u\} =$$

$$= \int_{A_u} P\{\max_{2^n \nmid k} \| W(\tfrac{k}{2^n}, \cdot)\| \le u \,|\, W(1,y)=f(y)\} d_f P\{W(1,y)\equiv f(y)\}$$

where the integral is taken over the set $A_u \subset C(0,1)$ with respect to the Wiener measure and $2^n \nmid k$ means that $2^n$ is not a divisor of $k$ $(1\le k\le 2^n)$ what simply means $k\ne 2^n$. Hence by Lemma 2 we have

$$P\{\max_{1\le k\le 2^n} \| W(\tfrac{k}{2^n}, \cdot)\| \le u\} \le P\{\max_{2^n \nmid k} \| W(\tfrac{k}{2^n}, \cdot)\| \le u \,|\, W(1,y)\equiv 0\} P(A_u) =$$

$$= P\{\max_{2^n \nmid k} \| W(\tfrac{k}{2^n}, \cdot)\| \le u \,|\, W(1,y)\equiv 0\} \; P\{\| W(\cdot)\| \le u\} .$$

Similarly by Lemmas 2 and 3 we have

$$P\{\max_{2^n \nmid k} \| W(\tfrac{k}{2^n}, \cdot)\| \le u \,|\, W(1,y)\equiv 0\} =$$

$$= \int_{A_u} P\{\max_{2^{n-1} \nmid k} \| W(\tfrac{k}{2^n}, \cdot)\| \le u \,|\, W(\tfrac{1}{2},y)=f(y), W(1,y)\equiv 0\} d_f P\{W(\tfrac{1}{2},y)\equiv f(y) \,|\, W(1,y)$$

$$\le P\{\max_{2^{n-1} \nmid k} \| W(\tfrac{k}{2^n}, \cdot)\| \le u \,|\, W(\tfrac{1}{2},y) \,|\, W(1,y)\equiv 0\} \; P\{\| W(\tfrac{1}{2}, \cdot)\| \le u \,|\, W(1,y)\equiv 0\} =$$

$$= P\{\max_{2^{n-1} \nmid k} \| W(\tfrac{k}{2^n}, \cdot)\| \le u \,|\, W(\tfrac{1}{2},y)\equiv W(1,y)\equiv 0\} P\{ \tfrac{1}{2} \| W(\cdot)\| \le u \}.$$

The next step of our proof goes as follows:

$$P\{\max_{2^{n-1} \nmid k} \| W(\tfrac{k}{2^n}, \cdot)\| \le u \,|\, W(\tfrac{1}{2}, y)\equiv W(1,y)\equiv 0\} =$$

$$= \int_{A_u \times A_u} P\{\max_{2^{n-2} \nmid k} \| W(\tfrac{k}{2^n}, \cdot)\| \le u \,|\, W(\tfrac{1}{4}, \cdot)=f_1, W(\tfrac{3}{4}, \cdot)=f_2, W(\tfrac{1}{2}, \cdot) = W(1,\cdot) = 0\}$$

$$d_{f_1,f_2} P\{W(\tfrac{1}{4},\cdot)=f_1,\ W(\tfrac{3}{4},\cdot)=f_2 \mid W(\tfrac{1}{2},\cdot)=W(1,\cdot)=0\} \le$$

$$\le P\{\max_{2^{n-2}\nmid k} \| W(\tfrac{k}{2^n},\cdot)\| \le u \mid W(\tfrac{j}{4},\cdot)=0;\quad j=1,2,3,4\}\ P\{\| W(\tfrac{1}{4},\cdot)\| \le u \mid W(\tfrac{1}{2},\cdot)=0\}$$

$$P\{\| W(\tfrac{3}{4},\cdot)\| \le u \mid W(\tfrac{1}{2},\cdot)=W(1,\cdot)=0\} =$$

$$= \prod_{k=0}^{3} P\{\max_{k2^{n-2}<j<(k+1)2^{n-2}} \| W(\tfrac{j}{2^n},\cdot)\| \le u \mid W(\tfrac{k}{4},\cdot)=W(\tfrac{k+1}{4},\cdot)=0\}\ .$$

$$(P\{8^{-1/2}\ \| W(\cdot)\| \le u\})^2$$

Continuing this procedure we get the statement of Lemma 4.

Now we are in the position to prove the second inequality of Theorem 1.

LEMMA 5. <u>There exist positive constants</u> $C_3, C_4$ <u>such that</u>

$$P\{M\le u\} \le C_3 \exp\{-C_4 u^{-2} \log 1/u\}\ .$$

PROOF. In Lemma 4 choose $n=[C \log u^{-1}]$ with a suitable $C>0$ and apply Consequence B.2 then one gets Lemma 5.

The proof of the other inequality of Theorem 1 is based on an approximation of the Wiener sheet.

Let

$$\begin{array}{l}
W_{01}(y),\\
W_{11}(y),\\
W_{21}(y),\ W_{22}(y)\\
\cdots\cdots\cdots\cdots\\
\cdots\cdots\cdots\cdots\\
\cdots\cdots\cdots\cdots\\
W_{n1}(y),\ W_{n2}(y),\ W_{n,2^{n-1}}(y),\\
\cdots\cdots\cdots\cdots\cdots\cdots\\
\cdots\cdots\cdots\cdots\cdots\cdots\\
\cdots\cdots\cdots\cdots\cdots\cdots
\end{array}$$

be a double array of independent Wiener processes. Define a sequence of Gaussian processes as follows:

$$\tilde{W}(1,y)=W_{o1}(y),$$

$$\tilde{W}(\tfrac{1}{2},y) = \frac{W(1,y)}{2} + \frac{W_{11}(y)}{\sqrt{4}}$$

.........................

.........................

.........................

$$\tilde{W}(\frac{2k+1}{2^n},y)= \frac{\tilde{W}(\frac{2k}{2^n},y)+\tilde{W}(\frac{2k+2}{2^n},y)}{2} + \frac{W_{n,(k+1)}(y)}{\sqrt{2^{n+1}}} .$$

......................... $(k=0,1,2,\ldots,2^{n-1}-1)$

.........................

.........................

and define the sequence

$$W_n(x,y) = \tilde{W}([2^n x]2^{-n},y) .$$

It is easy to see that the just defined sequence $\{W_n(x,y)\}$ is tending to a Wiener sheet. In fact we have

LEMMA A. (Csörgő-Révész, 1980) _The sequence_ $\{W_n(x,y)\}$ _is uniformly convergent on the unit square_ $I^2=[0,1]^2$ _with probability_ 1 _and its limit is a Wiener sheet_ $W(x,y)$ _for which_

$$W(\frac{k}{2^n},y) = \tilde{W}(\frac{k}{2^n}, y) .$$

This Lemma trivially implies

LEMMA 6. _We have_

$$\sup_{1\le k\le 2^n}\ \sup_{0\le y\le 1} |W(\tfrac{k}{2^n},y)| \le \sup_{0\le y\le 1}|W_{o1}(y)| +$$

$$+ \sum_{j=1}^{n} \sup_{1\le k\le 2^n} \frac{1}{\sqrt{2^{j+1}}} \sup_{0\le y\le 1}|W_{jk}(y)|$$

<u>and hence</u>

$$P\{\sup_{1\le k\le 2^n} \| W(\tfrac{k}{2n},\ \cdot)\| \le u\} \ge$$

$$\ge P\{\| W(y)\| \le c_o u\} \prod_{j=1}^{n} (P\{\| W\| \le c_j u\sqrt{2^{j+1}}\})^{2^j}$$

<u>where</u> $\{c_j\}$ <u>is a decreasing sequence of positive numbers with</u>

$$c_j\sqrt{2^{j+1}} \nearrow \infty\ ,\quad c_j \approx \frac{6}{\pi^2 j^2}\ ,\quad \sum_{j=o}^{\infty} c_j = 1\ .$$

Now, we are in the position to prove the first inequality of Theorem 1.

LEMMA 7. <u>There exist positive constants</u> $C_1, C_2$ <u>such that</u>

$$P(M\le u) \ge C_1 \exp(-C_2 u^{-2}(\log u^{-1})^5)\ .$$

PROOF. Assume (without loss of generality ) that $u<1$ . Let $\{c_j\}$ be the sequence of Lemma 6 and define the integers $n_1$ and $n_2$ as follows:

let $n_1$ be the largest integer for which

$$c_{n_1}\sqrt{2^{n_1+1}}\ u \le 4\ ,$$

let $n_2$ be the smallest integer for which

$$\exp\{-c_{n_2}^2\ u^2\ 2^{n_2}\} < \frac{1}{4}\, c_{n_2}\ u\sqrt{\frac{\pi}{2^{n_2}}}\ .$$

Then a simple calculation gives that $n_1<n_2$ , $n_2=O(\log u^{-1})$, $n_1=O(\log u^{-1})$ and hence $n_2-n_1=O(\log u^{-1})$ . Now, let $n>n_2$ and consider

$$P\{\sup_{1\le k\le 2^n} \| W(\tfrac{k}{2^n}, \cdot)\| \le u\} \ge$$

$$\ge P\{\|W\| \le c_o u\} \prod_{j=1}^{n_1} (P\{\|W\| \le c_j u\sqrt{2^{j+1}}\})^{2^j} \prod_{j=n_1+1}^{n_2} \prod_{j=n_2+1}^{n} =$$

$$= I_o\, I_1\, I_2\, I_3\ .$$

In order to get a lower estimation

for $I_o$ use Consequence B.2,

for $I_1$ use also Consequence B.2 and the estimation $n_1=O(\log u^{-1})$ ,

for $I_2$ use the inequality

$$P\{\|W\| \le c_j\sqrt{2^{j+1}}\, u\} \ge \varepsilon > 0$$

with a suitable fixed positive $\varepsilon$ for all $n_1<j\le n_2$ and the estimation $n_2-n_1=O(\log u^{-1})$ ,

for $I_3$ use Consequence B.1.

In this way for any integer $n$ we get the estimation

$$P\{\sup_{1\le k\le 2^n} \| W(\tfrac{k}{2^n}, \cdot)\| \le u\} \ge C_1 \exp(-C_2 u^{-2}(\log u^{-1})^5)$$

what implies our Lemma 7.

Hence Theorem 1 is proved.

## 3. STRONG THEOREMS

In this Section we fomulate the analgues of Consequences A.1 and A.2 for a Wiener sheet. We do not intend to give the proofs, since they can be obtained using standard methods only (see for example [2]). Here we use again the

DEFINITION. A function $f(x,y)$ is said to be $\varepsilon$ -constant on the rectangle $R=[x_1,x_2]x[y_1,y_2]$ if

$$\sup_{(x,y)\in R} |f(x,y) - f(x_1,y_2)| \le \varepsilon \quad .$$

(Here and in what follows a rectangle means a rectangle of the form $[x_1,x_2]x[y_1,y_2]$ .)

THEOREM 2. For all $T$ big enough and for almost all sample function of $W(x,y)$ there exists a random square $S\subset[O,T]x[O,T]$ of size $C\frac{\log T}{(\log\log T)^5}$ where $W(x,y)$ is $\varepsilon$ -constant. Here the constant $C=C(\varepsilon)$ depends only on $\varepsilon$ .

THEOREM 3. For all $T$ big enough, and for almost all sample function of $W(x,y)$ there is not any rectangle $R\in[O,T]x[O,T]$ of size $C\frac{\log T}{\log\log T}$ , where $W$ where $\varepsilon$ -constant. Here the $C=C(\varepsilon)$ depends only on $\varepsilon$ .

THEOREM 4. There exist positive constants $C_1,C_2$ such that

$$\liminf_{T\to\infty} \sup_{\substack{o\le x\le T\\ o\le y\le T}} \left(\frac{\log\log T}{T^2(\log\log\log T)^5}\right)^{1/2}|W(x,y)|\le C_2 \qquad \text{a.s.}$$

and

$$\liminf_{T\to\infty} \sup_{\substack{o\le x\le T\\ o\le y\le T}} \left(\frac{\log\log T}{T^2\log\log\log T}\right)^{1/2}|W(x,y)|\ge C_1 \qquad \text{a.s.}$$

REFERNCES

1 CSÖRGŐ, M.-RÉVÉSZ,P. (1979) How small are the increments of a Wiener process? Stoch.Proc. and Appl. 8, 119-129.

2 CSÖRGŐ, M.-RÉVÉSZ, P. (1978) How big are the increments of a multi-parameter Wiener process? Z. Wahrscheinlichkeitstheorie verw. Gebiete 42, 1-12.

3 CHUNG, K.L. (1948) On the maximum partial sums of sequences of independent random variables. Trans.Amer.Math.Soc. 64, 205-233.

4 GOODMAN, V. (1976) Distribution estimates for functionals of the two-parameter Wiener process. Ann. Probability 4, 977-982.

5 CSÖRGŐ, M.-RÉVÉSZ, P.(1980) Strong approximations in probability and statistics. Akadémiai Kiadó, Budapest.

# AN APPROACH TO THE FORMULA $L = \lambda V$ VIA THE THEORY OF STATIONARY POINT PROCESSES ON A SPACE OF COMPACT SUBSETS OF $R^k$

Tomasz Rolski

University of Wrocław

## 1. Introduction

Much attention has been drawn to the queueing formula $L = \lambda V$; see [1],[2],[3],[5],[10],[11],[12]. In this context L means "average queue", $\lambda$ means "intensity of arrivals" and V means "average delay in queue". In this note we point out that the formula $L = \lambda V$ is valid for stationary processes of random compact sets $\underline{\mu}$ in $R^k$. Informally speaking it is a collection of random compact sets strewed over $R^k$ in a stationary manner. A formal definition will be given in Section 4. In this context L means "average number of sets covering zero", $\lambda$ means intensity of $\underline{\mu}$, and V means "average volume of a typical set". Clearly concepts of the intensity and average volume of a typical set need explanations and they are studied in Sections 5 and 6. The formula $L = \lambda V$ is studied in Section 6. The proof of $L = \lambda V$ is based on the behaviour of sample path and applies Nguyen-Zessin's ergodic theorem for point processes; see [8]. Finaly in Section 7 we deal with the number of sets from a stationary process of random compact sets $\underline{\mu}$ overlapping a compact set K. The special case of such a problem was considered in [6], and recently, in this setting, was independently studied by Stoyan [12].

## 2. Preliminaries

Throughout the paper we denote the k-dimensional Euclidean space by $R^k$ and the Borel $\sigma$-field of subsets of $R^k$ by $\mathcal{B}^k$. The set of all non-negative real numbers we denote by $R_+$. For a topological spece $E$ we denote by $\mathcal{B}E$ the Borel $\sigma$-field of subsets of $E$. The open ball in a metric space $E$ with the center x and the radius r we denote by $B(E,x,r)$. The complement of a set A is $A^c$. The indicator function of a set A is $1_A(x)$. The following notations and convention are used:

$$(0,1)^k = (0,1) \times \ldots \times (0,1) \in \mathcal{B}^k ,$$

$$|B| = \int_B dx , \qquad B \in \mathcal{B}^k ,$$

$$\sum_i = \sum_{i=0}^{\infty} .$$

## 3. Random elements associated with point process

We follow Kallenberg [4] for the definition of random measures or point processes. Let $E$ be a locally compact second countability Hausdorff topological space (LCS). Let $M(E)$ be the class of measures $\mu$ such that $\mu(V) < \infty$, for any bounded V and $C(E)$ be the class of continuous functions f with compact support. We can make $M(E)$ into a Polish space with convergence $\mu_n \to \mu$ iff $\int f d\mu_n \to \int f d\mu$ for all $f \in C(E)$. The subset $M(E)$ of integer valued measures in $M(E)$ is Borel.

Let $(\Omega, F, \mathbb{P})$ be a probability space.

Definition 3.1:

A measurable mapping

$$\underline{\mu}: (\Omega, F) \to (M(E), \mathcal{B}M(E))$$

is said to be a random neasure on $E$. If $\Omega = M(E)$, $F = \mathcal{B}M(E)$ and $\underline{\mu}(\omega) = \omega$ then $\underline{\mu}$ is called the canonical random measure on $E$. The canonical random measure identifies with the probability space $(M(E), \mathcal{B}M(E), \mathbb{P})$.

Definition 3.2:

A measurable mapping

$$\underline{\nu}: (\Omega, F) \to (N(E), \mathcal{B}N(E))$$

is said to be a point process on $E$. If $\Omega = N(E)$, $F = \mathcal{B}N(E)$ $\underline{\nu}(\omega) = \omega$ then $\underline{\nu}$ is called the canonical point process on $E$. The canonical point process identifies with the probability space $(N(E), \mathcal{B}N(E), \mathbb{P})$.

Consider a Polish space $A$ and a locally compact secondcountability group $E$. There is given on $A$ a family ot automorphisms $\{\sigma_x\}$; $x \in E$ fulfilling

(a) $\sigma_x \circ \sigma_y = \sigma_{x+y}$, $x, y \in E$ (group property),

(b) the mapping $A \times E \ni (\alpha, x) \to \sigma_x \alpha \in M$ is measurable.

On $M(E)$ we define a family of automorphisms $\{\tau_x\}$, $x \in E$ by

$$\tau_x \mu(B) = \mu(B+x), \qquad x \in E, \quad B \in \mathcal{B}E .$$

Denote for $(\alpha,\mu) \in A \times M(E)$

$$T_x(\alpha,\mu) = (\sigma_x \alpha, \tau_x \mu), \qquad x \in E .$$

Let $\underline{\alpha}$ be a random element on a probability space $(\Omega,\mathcal{F},\mathbb{P})$ assuming values at $A$. Let $\underline{\mu}$ be a random measure on $E$ defined on the probability space $(\Omega,\mathcal{F},\mathbb{P})$.

Definition 3.3:

The pair $(\underline{\alpha},\underline{\mu})$ is called the random element associated with random measure. The $(\underline{\alpha},\underline{\mu})$ is said to be stationary if

$$\mathbb{P} \circ (\underline{\alpha},\underline{\mu})^{-1} \circ T_x = \mathbb{P} \circ (\underline{\alpha},\underline{\mu})^{-1}$$

or

$$P \circ T_x = P, \qquad x \in E, \tag{3.1}$$

where $P = \mathbb{P} \circ (\underline{\alpha},\underline{\mu})^{-1}$ is the distribution of $(\underline{\alpha},\underline{\mu})$.
In the case when $\underline{\mu}$ is a point process we adopt the abreviation REAPP for $(\underline{\alpha},\underline{\mu})$.

We aim now to define the Palm distribution of $P$. It is a simply modification of the definition from [7], where the Palm distribution was defined not for the random elements associated with random measure but for the random measure. From now on $P$ is stationary that is (3.1) holds. Define a measure $\lambda(\cdot)$ on $(E,\mathcal{B}E)$ by

$$\lambda(B) = E\underline{\mu}(B), \qquad B \in \mathcal{B}E . \tag{3.2}$$

We assume that $\lambda(\cdot)$ is non-zero and $\lambda(B) < \infty$ for bounded $B$. Then $\lambda(\cdot)$ is a Haar measure on $(E,\mathcal{B}E)$. Let $g\colon E \to R_+$ be such that

$$\int_E g(x)\lambda(dx) = 1$$

and set for $F \in \mathcal{B}(A \times M(E))$

$$P^o(F) = \int_{A\times M(E)} \int_E g(x)\, 1_F \circ T_x(\alpha,\mu)\mu(dx)P(d\alpha \times d\mu) .$$

Clearly $P^o$ is a probability measure. We shall call it as the Palm distribution of $P$.

Theorem 3.1 (Mecke [7]):

The Palm distribution $P^o$ of P is the only probability on $A \times M(E)$ such that for all nonnegative $\mathcal{B}(A \times M(E) \times E)$ -measurable functions $z((\alpha,\mu),x)$

$$(3.3) \qquad \int_{A\times M(E)} \int_E z((\alpha,\mu),x)\mu(dx)P(d\alpha \times d\mu) =$$

$$= \int_{A\times M(E)} \int_E z(T_{-x}(\alpha,\mu),x)\lambda(dx)P^o(d\alpha \times d\mu) \ .$$

Let $(\underline{\alpha},\underline{\nu})$ be a stationary REAPP and suppose that $\underline{\nu}$ is a point process on $R^k$. In such a case $E = R^k$, $\lambda(dx) = \lambda dx$, where $\lambda$ is the intensity of $\underline{\nu}$ and dx is the Lebesgue measure on $R^k$. Denote

$$A \times N^o(R^k) = \{(\alpha,\mu) \in A \times N(R^k): \mu(\{0\}) \geq 1\} \ .$$

We have the following corollary.

Corollary:

If $P^o$ is the Palm distribution of a REAPP $(\underline{\alpha},\underline{\nu})$, where $\underline{\nu}$ is a point process on $R^k$ then

$$P^o(A \times N^o(R^k)) = 1 \ .$$

Proof. Set in (3.3)

$$z((\alpha,\mu),x) = \begin{cases} 1, & \mu(\{x\}) \geq 1, \quad x \in (0,1)^k, \\ 0, & \text{otherwise.} \end{cases}$$

Recall also the following definition.

Definition 3.4:

It is said that a point process $\underline{\nu}$ on $E$ is without multiple points if

$$P(\{\omega: \underline{\nu}(\omega)(\{x\}) \leq 1, \ x \in E\}) = 1 \ .$$

## 4. Processes of random sets.

Let $K$ be the space of all compact sets in $R^k$ endowed with the Hausdorff metric given by

$$h(K_1,K_2) = \max(\sup_{x\in K_2} \kappa(x,K_1), \ \sup_{x\in K_1} \kappa(x,K_2)) \ .$$

Here $\kappa(x,K)$ denotes the distance from $x$ to $K$ in the Euclidean metric. It is known (see e.g. [6]) that $(\mathcal{K},h)$ is LCS.

Let $u(K)$ and $\rho(K)$ denote the center and the radius of the ball circumscribing a compact set $K \subset R^k$. The ball circumscribing a compact set K is a ball of the smallest volume containing K. The following argument shows existence and uniqueness. Define the function $r(x) = \min\{t: B(R^k,x,t) \supset K\}$, which is clearly continuous. Thus there exists $x_0$ such that $r(x_0) = \min_{x \in K} r(x)$. If there exists another ball $B(R^k,y_0,r(x_0)) \supset K$, $x_0 \neq y_0$ then also $B(R^k,(x_0-y_0)/2,(r^2(x_0) - (\|x_0-y_0\|/2/^2)^{\frac{1}{2}}) \supset K$. However in such a case $(r^2(x_0) - (\|x_0-y_0\|/2)^2)^{\frac{1}{2}} < r(x_0)$ which is impossible.

The mappings $u\colon \mathcal{K} \to R^k$ and $\rho\colon \mathcal{K} \to R_+$ are continuous and

$$u(K + x) = u(K) + x , \tag{4.1}$$

$$\rho(K + x) = \rho(K), \quad x \in R^k, \quad K \in \mathcal{K}. \tag{4.2}$$

Consider a canonical point process $\underline{\mu}$ on $\mathcal{K}$, that is a probability space $(N(\mathcal{K}),\mathcal{BN}(\mathcal{K}),\mathbb{P})$. We assume that $\underline{\mu}$ is stationary. It was shown in [4], Lemma 2.3, that

$$\underline{\mu} = \sum_i 1_{\{\underline{K}_i\}} , \tag{4.3}$$

where $\underline{K}_i$, $i=0,1,\ldots$ is a sequence of r.e.'s assuming values at $\mathcal{K}$.

<u>Definition</u> 4.1:

The point process $\underline{\mu}$ is said to be a process of random sets.

Using representation (4.3) define random measures

$$\underline{\mu}_\rho = \sum_i \rho(\underline{K}_i)\, 1_{\{u(\underline{K}_i)\}} , \tag{4.4}$$

$$\underline{\mu}_1 = \sum_i |\underline{K}_i|\, 1_{\{u(\underline{K}_i)\}} \tag{4.5}$$

and the point process

$$\underline{\mu}^* = \sum_i 1_{\{u(\underline{K}_i)\}} . \tag{4.6}$$

The measurability of mappings (4.4)-(4.6) it follows from the measurability of $\rho(K)$, $|K|$, $u(K)$. It is easy to show that $\underline{\mu}_\rho$, $\underline{\mu}_1$, $\underline{\mu}^*$ are sationary.

## 5. Intensity of $\underline{\mu}$.

Consider a stationary ergodic process of random sets $\underline{\mu}$. Let

(5.1) $$\lambda = E\,\underline{\mu}^*((0,1)^k)$$

be the intensity of $\underline{\mu}^*$. We assume $\lambda < \infty$, and $\mu^*$ is without multiple points.

Definition 5.1.:

The value $\lambda$ we call the intensity of the process of random sets $\underline{\mu}$.

The following definition was proposed in [8],[9].

Definition 5.2.:

A system of open, bounded, convex sets $\{Q_R\}$, $R > 0$ is said to be regular if

(a) $Q_R \subset B(R^k,0,R)$, $R > 0$,

(b) there exists $k > 0$ and $R_0 > 0$ such that $|Q_R| > k|B(R^k,0,R)|$, $R > R_0$.

In the sequel $\{Q_R\}$ always denotes a regular system of sets. The following proposition justifies Definition 5.1. Recall representation $\underline{\mu} = \sum_i 1_{\underline{K}_i}$.

Proposition 5.1:

The limit

(5.2) $$\lim_{R\to\infty} \frac{\#\{i:\ \underline{K}_i \subset Q_R\}}{|Q_R|} = \lambda\ , \qquad \text{a.s. } \mathbb{P}.$$

Moreover

(5.3) $$\lim_{R\to\infty} \frac{\#\{i:\ \underline{K}_i \cap Q_R \neq \emptyset\}}{|Q_R|} = \lambda, \qquad \text{a.s. } \mathbb{P}.$$

Proof. First we note that

$$\#\{i:\ \underline{K}_i \subset Q_R\}\ ,$$

$$\#\{i:\ u(\underline{K}_i) \in Q_R\}$$

are random variables. It follows from the fact that

$$C = \{B \in \mathcal{K}:\ B \subset Q_R\} \in \mathcal{BK} \qquad \text{(see [6])},$$

$$\#\{i: \underline{K}_i(\mu) \subset Q_R\} = \mu(C) ,$$

and that the mapping $\mu \to \mu(C)$ is measurable. Also $\#\{i: u(\underline{K}_i) \in Q_R\}$ is a random variable because $\underline{\mu}^*$ is a point process and

$$\#\{i: u(\underline{K}_i)(\mu) \in Q_R\} = \underline{\mu}^*(\mu)(Q_R) . \tag{5.4}$$

To prove (5.2) it suffices to show

$$\lim_{R\to\infty} \frac{\#\{i: u(\underline{K}_i) \in Q_R\}}{|Q_R|} = \lambda , \qquad \text{a.s. } \mathbb{P} \tag{5.5}$$

and

$$\lim_{R\to\infty} \left(\frac{\#\{i: u(\underline{K}_i) \in Q_R\}}{|Q_R|} - \frac{\#\{i: \underline{K}_i \subset Q_R\}}{|Q_R|}\right) = 0 , \quad \text{a.s. } \mathbb{P}. \tag{5.6}$$

The existence of the limit in (5.5) is the immediate consequence of (5.4) and the Corollary 1 after Theorem 1 in [7].

To prove (5.6) define a subset of $R^k$

$$J^-(Q,r) = \{x \in Q: B(R^k,x,r) \subset Q\} ,$$

where $Q \in \mathcal{B}^k$. Notice that $J^-(Q,r_1) \subset J^-(Q,r_2)$ whenever $r_1 \geq r_2$. We have for each $r > 0$

$$\begin{aligned}
&\{i: u(\underline{K}_i) \in Q_R\} - \{i: \underline{K}_i \subset Q_R\} \subset \\
&\subset \{i: u(\underline{K}_i) \in Q_R - J^-(Q_R,\rho(\underline{K}_i))\} = \\
&= \{i: u(K_i) \in Q_R - J^-(Q_R,\rho(\underline{K}_i)),\ 0 \leq \rho(\underline{K}_i) \leq r\} \cup \\
&\cup \{i: u(\underline{K}_i) \in Q_R - J^-(Q_R,\rho(\underline{K}_i)), \rho(\underline{K}_i) > r\} \subset \\
&\subset \{i: u(\underline{K}_i) \in Q_R - J^-(Q_R,r)\} \cup \{i: u(\underline{K}_i) \in Q_R, \rho(\underline{K}_i) > r\} .
\end{aligned}$$

From Lemma 2 in [8] we have

$$\lim_{R\to\infty} \frac{|Q_R - J^-(Q_R,r)|}{|Q_R|} = 0 .$$

Hence by the Corollary 1 after Theorem 1 from [8]

$$\lim_{R\to\infty} \frac{\#\{i:\ u(\underline{K}_i) \in Q_R - J^-(Q_R,r)\}}{|Q_R|} = 0 , \qquad \text{a.s. } \mathbb{P}.$$

The same corollary from [8] also provides

$$\lim_{R\to\infty} \frac{\#\{i:\ u(\underline{K}_i) \in Q_R, \rho(\underline{K}_i) > r\}}{|Q_R|}$$

$$= E(\#\{i:\ u(\underline{K}_i) \in (0,1)^k, \rho(\underline{K}_i) > r\}), \qquad \text{a.s. } \mathbb{P}.$$

Thus for completing the proof of (5.6) we remark that

$$\lim_{R\to\infty} E(\#\{i:\ u(\underline{K}_i) \in (0,1)^k, \rho(\underline{K}_i) > r\}) = 0 , \qquad \text{a.s. } \mathbb{P}$$

which follows from that

$$\#\{i:\ u(\underline{K}_i) \in (0,1)^k, \rho(\underline{K}_i) > r_1\} \le$$

$$\le \#\{i:\ u(\underline{K}_i) \in (0,1)^k, \rho(\underline{K}_i) > r_2\} \qquad \text{if } r_1 \ge r_2$$

and

$$\lim_{r\to\infty} E\#\{i:\ u(\underline{K}_i) \in (0,1)^k, \rho(\underline{K}_i) > r\}$$

$$= \lim_{r\to\infty} \lambda P^o_\rho(\{\mu:\ \mu(\{0\}) > r\}) = 0$$

where $P^o$ denotes the Palm distribution of the distribution of $\underline{\mu}$. This completes the proof of (5.2).

Now we prove (5.3). In view of (5.2) we need only to show

$$\text{(5.7)} \qquad \lim_{R\to\infty} \left(\frac{\#\{i:\ \underline{K}_i \cap Q_R \neq \emptyset\}}{|Q_R|} - \frac{\#\{i:\ \underline{K}_i \subset Q_R\}}{|Q_R|}\right) = 0 , \qquad \text{a.s. } \mathbb{P}.$$

However

$$\{i:\ \underline{K}_i \cap Q_R \neq \emptyset\} - \{i:\ \underline{K}_i \subset Q_R\} \subset$$

$$\subset \{i:\ u(\underline{K}_i) \in Q_R, \underline{K}_i \cap Q_R^c \neq \emptyset\} \cup \{i:\ u(\underline{K}_i) \notin Q_R, \underline{K}_i \cap Q_R \neq \emptyset\} \subset$$

$$\subset \{i:\ u(\underline{K}_i) \in Q_R - J^-(Q_R, \rho(\underline{K}_i))\} \cup$$

$$\cup \{i:\ u(\underline{K}_i) \in J^+(Q_R, \rho(\underline{K}_i)) - Q_R\} ,$$

where

$$J^{+}(Q,r) = \{x\colon B(R^{k},x,r) \cap Q \neq \emptyset\} .$$

It was shown in the proof of (5.2) that

$$\lim_{R\to\infty} \frac{\#\{i\colon u(\underline{K}_i) \in Q_R - J^{-}(Q_R,\rho(\underline{K}_i))\}}{|Q_R|} = 0 , \quad \text{a.s. } \mathbb{P}.$$

In the similar way we can show that

$$\lim_{R\to\infty} \frac{\#\{i\colon u(\underline{K}_i) \in J^{+}(Q_R,\rho(\underline{K}_i)) - Q_R\}}{|Q_R|} = 0 , \quad \text{a.s. } \mathbb{P}.$$

which completes the proof of (5.3).

Remark:

We related with each $K_i$ the point $u(K_i)$ from $B(R^k,u(K_i),\rho(K_i))$. Sometimes it is more convenient to choose not $u(K_i)$ but another point $u'(K_i) \in B(R^k,u(K_i),\rho(K_i))$. This is just the case in queueing theory where $K_i$ are intervals in $R^1$ and $u'(K_i)$ is the left end-point of $K_i$. We assume that the function $u'$ is measurable. Then $\underline{\mu}' = \sum_i 1_{\{u'(\underline{K}_i)\}}$ is a point process and

$$\text{(5.8)} \qquad \lim_{R\to\infty} \frac{\underline{\mu}'(Q_R)}{|Q_R|} = \lambda , \quad \text{a.s. } P.$$

It follows from the inequalities

$$\#\{i\colon \underline{K}_i \subset Q_R\} \le \underline{\mu}'(Q_R) \le \#\{i\colon B(R^k,u(\underline{K}_i),\rho(\underline{K}_i)) \cap Q_R \neq \emptyset\}$$

and

$$\text{(5.9)} \qquad \lim_{R\to\infty}\left(\frac{\#\{i\colon B(R^k,u(\underline{K}_i),\rho(\underline{K}_i)) \cap Q_R \neq \emptyset\}}{|Q_R|} - \frac{\#\{i\colon \underline{K}_i \subset Q_R\}}{|Q_R|}\right) = 0$$

$$\text{a.s. } \mathbb{P}.$$

The proof of (5.9) is similar to the proof of (5.7). Note also that $\underline{\mu}$ need not to be stationary unless $u'(K) + x = u'(K+x)$, $x \in R^k$, $K \in \mathcal{K}$.

## 6. Formula $L = \lambda V$.

Let $\underline{\mu}$ be a canonical process of random sets in $R^k$. In other words $\underline{\mu}$ is a point process on $\mathcal{K}$ and has representation $\underline{\mu} = \sum_i 1_{\{\underline{K}_i\}}$.

Define $\underline{L}(t): N(K) \to R_+$ by

$$\underline{L}(t) = \sum_i 1_{\{\underline{K}_i\}}(t) , \qquad t \in R^k ,$$

i.e. $\underline{L}(t)$ is the number of sets from $\{\underline{K}_i\}$ covering the point t. Let $\{\sigma_s\}$, $s \in R^k$ be a group of automorphisms on the class of functions $f: R^k \to R^1$ defined by

$$\sigma_s f(t) = f(t-s) , \qquad t,s \in R^k .$$

Lemma 6.1:

$\{L(t)\}$, $t \in R^k$ is a random process on $(N(K), BN(K), \mathbb{P})$. Moreover if $\mathbb{P}$ is stationary then $\{\underline{L}(t)\}$ is.

Proof. Measurability:

$$\{\mu: \underline{L}(s)(\mu) = k\} = \{\mu: \mu(K_{\{s\}}) = k\} ,$$

where as it was shown in [6]

$$K_{\{s\}} = \{K \in K: K \cap \{s\} \neq \emptyset\} \in BK .$$

For the stationarity it suffices to point out that

$$\sigma_t \underline{L}(s)(\mu) = \underline{L}(s)(\tau_t \mu) , \qquad t,s \in R^k, \quad \mu \in N(K) .$$

Consider a REAPP $(\underline{\mu}_1, \underline{\mu}^*)$, where $\underline{\mu}_1$ and $\underline{\mu}^*$ are defined in (4.5) and (4.6). Let $\mathbb{P}_1$ be the distribution $(\underline{\mu}_1, \underline{\mu}^*)$ and $\mathbb{P}_1^o$ its Palm distribution. Denote $P_1^o(A) = \mathbb{P}_1^o(A \times N(R^k))$, $A \in R^k$. The measure $\lambda(\cdot)$ defined in (3.2) is $\lambda dx$, where $\lambda$ is the intensity of $\underline{\mu}^*$ and dx is the Lebesgue measure in $R^k$. Then for a non-negative $B(M(R^k) \times R^k)$ - measurable function $z(\mu,x)$ we have by (3.3)

$$(6.1) \qquad \int_{M(R^k) \times N(R^k)} \int_{R^k} z(\mu,x)\nu(dx)\mathbb{P}_1(d\mu \times d\nu)$$

$$= \lambda \int_{M(R^k) \times R^k} \int_{R^k} z(\tau_{-x}\mu,x)dx \, \mathbb{P}_1^o(d\mu \times d\nu) .$$

Lemma 6.2:

$$E\, \underline{\mu}_1((0,1)^k) = \lambda\, E_{P_1^o}\, \underline{\mu}_1(\{0\}) ,$$

where

$$E_{P_1^o} \underline{\mu}_1(\{0\}) = \int_{M(R^k)} \mu(\{0\}) P_1^o(d\mu) .$$

Proof. Substitute to (6.1) the function

$$z(\mu,x) = \begin{cases} \mu(\{x\}) , & x \in (0,1)^k , \\ 0 , & \text{otherwise} , \end{cases}$$

Then

$$z(\tau_{-x}\mu,x) = \begin{cases} \mu(\{0\}) , & x \in (0,1)^k , \\ 0 , & \text{otherwise.} \end{cases}$$

Theorem 6.1:

If $\underline{\mu}$ is a stationary process of random sets with the finite intensity $\lambda$ then

$$(6.2) \qquad E\underline{L}(t) = \lambda E_{\mathbb{P}_1^o} |\underline{K}| ,$$

where

$$E_{\mathbb{P}_1^o} |\underline{K}| = \int_{M(R^k)} \mu(\{0\}) P_1^o(d\mu) .$$

Proof. Denote

$$a(R) = \{i: u(\underline{K}_i) \in Q_R\} , \qquad b(R) = \{i: \underline{K}_i \subset Q_R\} ,$$

$$c(R) = \{i: \underline{K}_i \cap Q_R \neq \emptyset\} .$$

We have

$$\sum_{i \in a(R)} |\underline{K}_i| = \underline{\mu}_1(Q_R)$$

and from the Corollary 1 after Theorem 1 in [8]

$$\lim_{R\to\infty} \frac{\underline{\mu}_1(Q_R)}{|Q_R|} = E\, \underline{\mu}_1((0,1)^k) | J) , \qquad \text{a.s. } \mathbb{P},$$

where $J$ denotes the $\sigma$-field of invariant sets with respect to $\{\tau_x\}$ on $(N(K), BN(K), \mathbb{P})$. Denote

$$\underline{\Psi} = E(\underline{\mu}_1((0,1^k))|J) .$$

From Lemma 6.1 we have

$$E\underline{\Psi} = \lambda \int_{M(K)} \mu(\{0\}) P_1^o(d\mu)$$

The proof of the theorem is based on the inequality

$$(6.3) \qquad \sum_{i \in b(R)} |\underline{K}_i| \leq \int_{Q_R} \underline{L}(s)ds \leq \sum_{i \in c(R)} |\underline{K}_i| .$$

Thus it suffices to show that

$$(6.4) \qquad \lim_{R\to\infty} \frac{\sum_{i\in b(R)} |\underline{K}_i|}{|Q_R|} = \underline{\Psi} , \qquad \text{a.s. } \mathbb{P},$$

$$(6.5) \qquad \lim_{R\to\infty} \frac{\sum_{i\in c(R)} |\underline{K}_i|}{|Q_R|} = \underline{\Psi} , \qquad \text{a.s. } \mathbb{P}.$$

To prove (6.4) we find by a similar argument as in the proof of Proposition 5.1 that for each $r > 0$

$$0 \leq \sum_{i\in a(R)} |\underline{K}_i| - \sum_{i\in b(R)} |K_i| \leq$$

$$\leq \sum_{\{i:u(\underline{K}_i)\in Q_R - J^-(Q_R,r)\}} |\underline{K}_i| + \sum_{\{i:u(\underline{K}_i)\in Q_R, \rho(K_i)>r\}} |\underline{K}_i| .$$

Lemma 2 and the Corollary 1 after Theorem 1 in [8] yields

$$\lim_{R\to\infty} \frac{\sum_{\{i:u(\underline{K}_i)\in Q_R - J^-(Q_R,r)\}} |K_i|}{|Q_R|}$$

$$= \lim_{R\to\infty} \frac{\underline{\mu}_1(Q_R - J^-(Q_R,r))}{|Q_R|} = 0 , \qquad \text{a.s. } \mathbb{P}.$$

They also provide existence of the limit

$$\lim_{R\to\infty} \frac{\sum_{\{i:u(K_i)\in Q_R,\rho(K_i)>r\}} |K_i|}{|Q_R|} =$$

$$= \lim_{R\to\infty} \frac{\sum_{i\in a(R)} |\underline{K}_i| 1_{(r,\infty)}(\rho(\underline{K}_i))}{|Q_R|} = \underline{\Phi}_r , \qquad \text{a.s. } \mathbb{P}.$$

We have $\underline{\Phi}_{r_1} \geq \underline{\Phi}_{r_2}$, whenever $r_1 \leq r_2$ and $\lim_{r\to\infty} E\underline{\Phi}_r = 0$ which yields $\lim_{r\to\infty} \underline{\Phi}_r = 0$, a.s.$\mathbb{P}$. The proof of (6.5) is similar. Thus by (6.3)-(6.5),

$$\lim_{R\to\infty} \frac{\int_{Q_R} \underline{L}(s)ds}{|Q_R|} = \underline{\Psi} , \qquad \text{a.s. } \mathbb{P}.$$

On the other hand from the stationarity of $\{\underline{L}(s)\}$, by the ergodic theorem

$$\lim_{R\to\infty} \frac{\int_{Q_R} \underline{L}(s)ds}{|Q_R|} = \underline{\mathbf{Y}} , \qquad \text{a.s. } \mathbb{P}$$

and

$$E\underline{Y} = E\underline{L}(0) .$$

Hence $\underline{Y} = \underline{\Psi}$ , a.s. $\mathbb{P}$ and

$$E\underline{Y} = E\underline{L}(t) = E\underline{\Psi} = \lambda E_{\mathbb{P}_1^o} |\underline{K}| .$$

## 7. The number of sets overlapping a set K

We finish the paper contributing to the Matheron's textbook [6], where a special process of random sets was derived (namely stationary Poisonian).

Let $\underline{\mu}$ be a stationary canonical process of random sets. Recall that $\underline{\mu} = \sum_i 1_{\{\underline{K}_i\}}$, where $\underline{K}_i: \mathcal{N}(\mathcal{K}) \to \mathcal{K}$ are measurable. For a fixed set $K \in \mathcal{K}$ define

$$D = \{F \in \mathcal{K}: F \cap K \neq \emptyset\} .$$

From [6] it follows that $D \in \mathcal{BK}$. We ask for the expected value

(7.1) $$E\#\{i: \underline{K}_i \cap K \neq \emptyset\} = E\underline{\mu}(D) .$$

Let $\mathbb{P}_*$ denotes the distribution of the REAPP $(\underline{\mu},\underline{\mu}^*)$ and $\mathbb{P}_*^o$ its Palm distribution. The marginal distribution we denote by $P_*^o(A) = \mathbb{P}_*^o(A \times N(R^k))$, $A \in BN(K)$. Assume that $0 < \lambda = E\underline{\mu}^*((0,1)^k) < \infty$ and that $\underline{\mu}^*$ is without multiple points.

To find $E\underline{\mu}(D)$ we use formula (3.3) setting for $\mu = \sum_i 1_{\{K_i\}} \in N(K)$

$$z(\mu,x) = \begin{cases} 1, & K_i \cap K \neq \emptyset, \quad u(K_i) = x, \\ 0, & \text{otherwise.} \end{cases}$$

Denote for $K,F \in K$

$$K \oplus \check{F} = \{x \in R^k: K \cap (F+x) \neq \emptyset\} .$$

From [6] it follows that the mapping $K \in F \to \check{F} \oplus K \in K$ is measurable. Note that

$$\int_{R^k} z(\mu,x)\underline{\mu}^*(\mu)(dx) = \underline{\mu}^*(\mu)(D) .$$

Consider the set

$$N^o(K) = \{\mu: \mu^*(\mu)(\{0\}) > 0\} \in BN(K)$$

Define

$$M_i = \{\mu: N^o(K): u(\underline{K}_i(\mu)) = 0\} , \qquad i=0,1,...$$

and

$$\underline{\ell}(\mu) = \begin{cases} i, & \text{if } \mu \in M_i , \\ 0, & \text{otherwise} . \end{cases}$$

Then $M_i$, $i=0,1,...$ , are measurable and

$$\underline{K}_{(0)}(\mu) = \underline{K}_{\underline{\ell}(\mu)}(\mu)$$

is a random element. We have for $\mu \in N^o(K)$

$$\int_{R^k} z(\tau_{-x},x)dx = |K \oplus K_0| .$$

Hence by (3.3)

$$E\underline{\mu}(D) = \int_{N(K)} \int_{R^k} z(\mu,x)\underline{\mu}^*(\mu)(dx)\mathbb{P}(d\mu) =$$

$$= \lambda \int_{N(K)} \int_{R^k} z(\tau_{-x}\mu,x)dx\, P^o_*(d\mu) = \lambda \int_K |K \oplus F| P^o_{(0)}(dF),$$

where $P^o_{(0)}$ is the distribution of random element $\underline{K}_{(0)}$ on $(N(K), BN(K), P^o_*)$. Thus we arrived at the relation

$$E\#\{i: \underline{K}_i \cap K \neq \emptyset\} = \lambda \int_{N(K)} |K \oplus \check{F}| P^o_{(0)}(dF) \; . \tag{7.2}$$

Note that if $K = \{0\}$ then (7.2) reduces to (6.2).

## References


[1] Brumelle, S.L.: On the relation between customer and time averages in queues, J.Appl.Probability, 8 (1971) p.508-520.

[2] ——— : A generalization of $L = \lambda W$ to moments of queue lenght and waiting times, Operat.Res., 20 (1972) p.1127-1136.

[3] Franken, P.: Einige Anwendunken der Theorie zufälliger Punktprozesse in der Bedienungstheorie I, Math.Nachr. 70 (1976) p.309-319.

[4] Kallenberg, O.: Random measures, Akademie-Verlag, Berlin, 1976.

[5] Little, J.D.C.: A proof for the queueing formula: $L = \lambda W$, Operat. Res., 9 (1961) p.383-387.

[6] Matheron, G.: Random sets and integral geometry, John Wiley & Sons, 1975.

[7] Mecke, J.: Stationäre zufällige Masse auf lokalkompakten Abelschen Gruppen, Z.Wahrscheinlichkeitstheorie und verw. Gebiete, 9 (1967) p.36-58.

[8] Nguyen, X.X., Zessin, H.: Punktprozesse mit Wechselwirkung, Z. Wahrscheinlichkeitstheorie verw. Gebite, 37 (1976) p.91-126.

[9] Pitt, H.R.: Some generalization of the ergodic theorem, Proc. Cambrigde Philos.Soc. 38 (1942) p.325-343.

[10] Stidham, S., Jr.: A last word on $L = \lambda W$, Operat.Res. 22 (1974) p.417-422.

[11] Stidhan, S., Jr.: On the relation between time averages and customer averages in stationary random marked point processes, NCSU-IE Technical Report No. 79-1.

[12] Stoyan, D.: Proof of some fundamental formulas of stereology without Poisson-type assumptions, Preprint, Bergakademie Freiberg, 1979.

# SEQUENTIAL ESTIMATES OF A REGRESSION FUNCTION BY ORTHOGONAL SERIES WITH APPLICATIONS IN DISCRIMINATION

Leszek Rutkowski

Technical University of Częstochowa, Poland

## 1. Introduction

Let $(X,Y)$ be a pair of random variables. X takes values in a Borel set A, $A\subset R^p$, whereas Y takes values in R. Let f be the marginal Lebesgue density of X. Based on a sample $(X_1,Y_1),\ldots,(X_n,Y_n)$ of independent observations of $(X,Y)$ we wish to estimate the regression r of Y on X, i.e.

$$r(x) = E[Y|X=x].$$

Let $\{g_k\}$, $k=0,1,2,\ldots$ be a complete orthonormal system defined on A, such that

$$|g_k(x)| \leqslant G_k \tag{1}$$

for all $x\in A$, where $\{G_k\}$ is a sequence of numbers. Define

$$h(x) = r(x)f(x).$$

We assume that functions h and f have the representations

$$h(x)\sim\sum_{k=0}^{\infty} a_k g_k(x), \tag{2}$$

$$f(x) \sim \sum_{k=0}^{\infty} b_k g_k(x) , \tag{3}$$

where

$$a_k = \int_A h(x) g_k(x)\, dx = E\left[Y g_k(X)\right] , \tag{4}$$

$$b_k = \int_A f(x) g_k(x)\, dx = E\left[g_k(X)\right] . \tag{5}$$

Let $\{N(n)\}$ be a sequence of integers, such that

$$\lim_{n \to \infty} N(n) = \infty \quad . \tag{6}$$

We propose a nonparametric estimate for the regression $r(x)$ as follows:

$$r_n(x) = h_n(x)/f_n(x) , \tag{7}$$

where

$$h_n(x) = \frac{1}{n} \sum_{i=1}^{n} \sum_{k=0}^{N(i)} Y_i g_k(X_i) g_k(x), \tag{8}$$

$$f_n(x) = \frac{1}{n} \sum_{i=1}^{n} \sum_{k=0}^{N(i)} g_k(X_i) g_k(x) . \tag{9}$$

Observe that estimates (8) and (9) may be expressed as

$$h_n(x) = \frac{n-1}{n} h_{n-1}(x) + \frac{1}{n} \sum_{k=0}^{N(n)} Y_n g_k(X_n) g_k(x),$$

$$f_n(x) = \frac{n-1}{n} f_{n-1}(x) + \frac{1}{n} \sum_{k=0}^{N(n)} g_k(X_n) g_k(x).$$

Thus, the unknown regression function $r(x)$ is estimated sequentially. It should be noted that estimates (9) and (7) are recursive versions of those introduced by Čencov [2] and Greblicki [5], respectively. Moreover, estimator (7) is analogous to a recursive version of kernel estimates studied by Greblicki [5], Ahmad and Lin [1], and Devroye and Wagner [4]. For surveys on commonly used nonparametric estimates of a probability density and a regression functions, the reader is referred to **Wertz and Schneider** [11] **and Devroye** [3].

In this paper we investigate weak and strong pointwise consistency of estimates (7). We derive from these estimates discrimination rules and establish their Bayes risk consistency.

## 2. Weak and Strong Pointwise Consistency

The consistency of estimate (7) is examined at points at which the density of X is positive. The weak pointwise consistency of $r_n$ is established in the following:

Theorem 1.
If $EY^2 < \infty$, conditions (1) and (6) are satisfied, and

$$\frac{1}{n^2} \sum_{i=1}^{n} \left( \sum_{k=0}^{N(i)} G_k^2 \right)^2 \xrightarrow{n} 0, \tag{10}$$

then

$$r_n(x) \xrightarrow{n} r(x) \quad \text{in probability}$$

at every point $x \in A$ at which series (2) and (3) converge to $h(x)$ and $f(x)$, respectively.

Proof. It suffices to show that

$$h_n(x) \xrightarrow{n} h(x) \quad \text{in probability}$$

and

$$f_n(x) \xrightarrow{n} f(x) \quad \text{in probability.}$$

We shall only prove the first assertion, the second can be proved similarly. It is clear that

$$h_n(x) - h(x) = \frac{1}{n} \sum_{i=1}^{n} Z_i(x) \quad + \quad (Eh_n(x) - h(x)),$$

where

$$Z_i(x) = \sum_{k=0}^{N(i)} (Y_i\, g_k(X_i) - a_k)\, g_k(x) \, .$$

From (4) and (8) it follows that

$$Eh_n(x) = \frac{1}{n} \sum_{i=1}^{n} \sum_{k=0}^{N(i)} a_k\, g_k(x).$$

Using well known properties of arithmetic means (see Tucker [10]), we deduce that estimator (8) is asymptotically unbiased at every point $x \in A$ at which

$$\sum_{k=0}^{N(n)} a_k\, g_k(x) \quad \xrightarrow{n} \quad h(x).$$

By Cauchy's inequality and (1) we have

$$E Z_n^2(x) \leqslant \sum_{k=0}^{N(n)} g_k^2(x) \sum_{k=0}^{N(n)} \operatorname{var} g_k(X_n) Y_n \leqslant EY^2 \left[\sum_{k=0}^{N(n)} G_k^2\right]^2 .$$

By virtue of the weak law of large numbers (see Loéve [7]) we complete the proof.

The strong pointwise consistency of $r_n$ is obtained in the following:

Theorem 2.
If $EY^2 < \infty$, conditions (1) and (6) are satisfied, and

$$\sum_{n=1}^{\infty} \frac{1}{n^2} \left[\sum_{k=0}^{N(n)} G_k^2\right]^2 < \infty \quad , \tag{11}$$

then

$$r_n(x) \xrightarrow{n} r(x) \quad \text{with probability one}$$

at every point $x \in A$ at which series (2) and (3) converge to $h(x)$ and $f(x)$, respectively.

Proof. Similar to Theorem 1 we shall only show that

$$h_n(x) \xrightarrow{n} h(x) \quad \text{with probability one.}$$

Applying Kolmogorov's second moment version of the strong law of large numbers (see Loéve [7]) one gets

$$\frac{1}{n} \sum_{i=1}^{n} Z_i(x) \xrightarrow{n} 0 \quad \text{with probability one.}$$

This implies the statement of the theorem.

Remark 1.

If $G_0 = \text{const}$, $G_k = \text{const}\ k^t$, $k = 1,2,\ldots$, conditions (10) and (11) take forms

$$\frac{1}{n^2} \sum_{i=1}^{n} [N(i)]^{4t+2} \xrightarrow{n} 0,$$

$$\sum_{n=1}^{\infty} \frac{1}{n^2} [N(n)]^{4t+2} < \infty .$$

For Hermite, Laguerre, Legendre and trigonometric orthonormal systems $t = -1/12$, $-1/4$, $1/2$, $0$, respectively (see Szegö [9]).

Remark 2.

Conditions for pointwise convergence of series (2) and (3) for different orthonormal systems can be found in Sansone [8].

3. An application in discrimination

Let $S = \{1,\ldots,M\}$; elements of S will be called classes. Let $(X,Y)$ be a pair of random variables; Y takes values in S and $P[Y = j] = p_j$ is the prior probability of the class j, X takes values in A, $A \subset R^p$, $f_j$ is the conditional density of X (with respect to the Lebesgue measure) given $Y = j$. $L(i,j)$ is the loss incurred in taking action $i \in S$ when the class is j. We assume 0 - 1 loss function. For a decision function $m: A \to S$ the expected loss is

$$R(m) = \sum_{i=1}^{M} p_i \int_A L(m(x), i)\, f_i(x)\, dx .$$

A decision function $m_0$ which classifies every $x \in A$ as coming from any class i for which

$$p_i f_i(x) = \max_j p_j f_j(x)$$

is a Bayes decision function. Define

$$t_j = \begin{cases} 1 \text{ for } Y = j \\ 0 \text{ for } Y \neq j \end{cases}$$

and

$$r_j(x) = E\left[t_j | X = x\right],$$

$$h_j(x) = r_j(x) f(x) = p_j f_j(x),$$

where $f(x) = \sum_{i=1}^{M} p_i f_i(x)$. A Bayes decision function can be rewritten in the following form: assign every $x \in A$ to any class i for which

$$h_i(x) = \max_j h_j(x).$$

Let $(X_1, Y_1), \ldots, (X_n, Y_n)$ be a sample of independent observations of $(X, Y)$. Define

$$t_i^j = \begin{cases} 1 \text{ for } Y_i = j \\ 0 \text{ for } Y_i \neq j . \end{cases}$$

Now we consider an empirical decision function $m_n$ which classifies every $x \in A$ to any class i for which

$$h_{in}(x) = \max_j h_{jn}(x),$$

where $h_{jn}(x)$ is an estimate of $h_j(x)$. As the estimate of $h_j(x)$ we take

$$h_{jn}(x) = \frac{1}{n} \sum_{i=1}^{n} \sum_{k=0}^{N(i)} t_i^j g_k(X_i) g_k(x).$$

By Theorems 1 and 2 one has

$$h_{in}(x) \xrightarrow{n} h_i(x)$$

in probability and with probability one, at every point $x \in A$ at which

$$\sum_{k=0}^{N(n)} a_k^i g_k(x) \xrightarrow{n} h_i(x), \tag{12}$$

where

$$a_k^i = \int_A h_i(x) g_k(x) dx .$$

Consequently, by Greblicki's results [6]

$$R(m_n) \xrightarrow{n} R(m_0)$$

weakly and strongly if (12) holds almost everywhere.

**Acknowledgment**

**The author wish to thank the referee whose suggestions led to improvements in the original manuscript.**

References

[1] Ahmad, I. A. and Lin, P. E., "Nonparametric sequential estimation of a multiple regression function," Bull. Math. Statist., vol. 17, pp. 63-75, 1976.

[2] Čencov, N. N., "Evaluation of an unknown distribution density from observations," Soviet Math., vol. 3, pp. 1559-1562, 1962.

[3] Devroye, L. P., "Universal consistency in nonparametric regression and nonparametric discrimination," Technical Report School of Computer Science, McGill University, 1978.

[4] Devroye, L. P. and Wagner, T. J., "On the L1 convergence of kernel estimators of regression functions with applications in discrimination," to appear in Z. Wahrscheinlichkeitstheorie und Verw. Gebiete.

[5] Greblicki, W., "Asymptotically optimal probabilistic algorithms for pattern recognition and identification," Scientific Papers of the Institute of Technical Cybernetics of Wrocław Technical University No. 18, Series: Monographs No. 3, Wrocław 1974.

[6] Greblicki, W., "Asymptotically optimal pattern recognition procedures with density estimates," IEEE Trans. Inform. Theory, vol. IT-24, pp. 250-251, 1978.

[7] Loéve, M., "Probability Theory I," 4th Edition, Springer-Verlag, 1977.

[8] Sansone, G., "Orthogonal functions," Interscience Publishers Inc., New York, 1959.

[9] Szegö, G., "Orthogonal polynomials," Amer. Math. Soc. Coll. Publ., vol. 23, 1959.

[10] Tucker, H. G., "A graduate course in probability," Academic Press, 1967.

[11] Wertz, W. and Schneider, B., "Statistical density estimation: a bibliography," Internat. Statist. Rev., vol. 47, pp. 155-175, 1979.

THE ASYMPTOTIC DISTRIBUTION OF CERTAIN GOODNESS OF FIT TEST STATISTICS

K.Sarkadi

Mathematical Institute of the Hungarian Academy of Sciences

1. Introduction. This paper deals with the asymptotic behaviour of statistics of the form

$$W_n = \frac{\left(\sum_{i=1}^{n} a_{in} X_{in}\right)^2}{\sum_{i=1}^{n} (X_i - \bar{X})^2} \tag{1}$$

where $X_1, \ldots, X_n$ are independent, identically distributed random variables with distribution function $F(x)$, $X_{1n} \leq \ldots \leq X_{nn}$ is their rearrangement in the order of magnitude, $\bar{X} = \sum_{i=1}^{n} X_i/n$ ; $a_{1n}, \ldots, a_{nn}$ are appropriately chosen constants. Conditions under which the asymptotic distribution of $W_n$ is normal are given.

Several tests of goodness of fit (Shapiro and Wilk [19,20], D'Agostino [3,4], De Wet and Venter [5], Filliben[7], Greenwood [9], Hahn and Shapiro [10], LaBrecque [11], Shapiro and Francia [18] Smith and Bain [23], Stephens [24], Sarkadi [17]) have the test statistics of the above type (with different definitions of $a_{in}$) or their test statistics can be equivalently written in this form (statistics which differ from each other in a strictly monotonic transformation determine one and the same test).

The properties of the mentioned tests have been studied mostly by Monte Carlo methods (see also Dyer [6], Shapiro, Wilk and Chen [21], Pearson, D'Agostino and Bowman [13]), the asymptotic distributions of the test statistic has been determined by D'Agostino [3] and DeWet and Venter [5] but even they considered in this respect the null hypothesis only. The present result opens the way of further investigation of the properties of these tests.

Let us remark that the consistency of some tests of the above type and of some related tests have been proved in Sarkadi [15,17] and Gerlach [8].

Here in Section 3. the inconsistency of some tests is proved.

2. The asymptotic distribution.

Throughout this Section we suppose that the fourth moment of $F(x)$ exists. We suppose that the first two moments are zero and one, respectively (the test statistics of the mentioned tests are location and scale invariant). The fourth moment will be denoted by $\beta_2$. Further, we suppose that $F'(x)=f(x)$ is positive and continuous in the interior of the support of the distribution.

Let be

$$\left.\begin{array}{l} Q_n(x)=X_{in} \\ J_n(x)=a_{in}\sqrt{n} \end{array}\right\} \quad \text{if} \quad \frac{i-1}{n}<x\leq\frac{i}{n}$$

$$(i=1,\ldots,n)$$

and suppose that the limit

$$\lim_{n\to\infty} J_n(x)=J(x)$$

exists almost everywhere in $0<x<1$.

Let us introduce, further, the following notations:

$$T_n = n^{-1/2} \sum_{i=1}^{n} a_{in} X_{in}$$

$$S_n = \sum_{i=1}^{n} (X_i - \bar{X})^2$$

$$T_{kn} = \frac{1}{k} \sum_{i=1}^{k} J\left(\frac{i}{k+1}\right) Q_n\left(\frac{i}{k+1}\right)$$

$$\mu = \int_0^1 J(x) F^{-1}(x)\,dx \qquad \mu_k = \frac{1}{k} \sum_{i=1}^{k} J\left(\frac{i}{k+1}\right) F^{-1}\left(\frac{i}{k+1}\right)$$

$$\sigma_o^2 = \int_0^1 \int_0^1 J(x)\, h(x) J(y) h(y) K(x,y)\,dx\,dy$$

$$K(x,y) = K(y,x) = x(1-y) \quad \text{for} \quad 0 \le x \le y \le 1$$

$$h(x) = 1/f(F^{-1}(x))$$

$$c = \int_0^1 J(x) h(x) \{x - \int_0^x [F^{-1}(y)]^2 dy\}\,dx$$

$$\sigma^2 = 4\mu^2 \sigma_o^2 - 4\mu^3 c + \mu^4 (\beta_2 - 1)$$

$F^{-1}(x)$ is the inverse of $F(x)$.

We suppose that the assumptions of Theorem 1. of Shorack [22] are fulfilled by $J_n(x)$, $J(x)$ and $F^{-1}(x) = g(x)$.

We prove the following theorem:

THEOREM 1. Under the above assumptions,

$$\lim_{n \to \infty} \mathcal{L}\left[n^{1/2}(W_n - \mu^2)\right] = \mathcal{N}(0, \sigma^2)$$

Proof. Using the method presented in Proposition 5. of Chernoff, Gastwirth and Johns [2] we derive the joint asymptotic distribution of $n^{1/2}(T_{kn}-\mu_k)$ and $n^{-1/2}(\sum_{i=1}^{n} X_i^2 -n)$ (a bivariate normal distribution).

Using Theorem 1. of Shorack [22] we can derive the asymptotic distribution of $n^{1/2}(T_{kn}-\mu_k-T_n+\mu)$. The second moment of this distribution turns out to tend to 0 with increasing $k$. Similarly, the second moment of $n^{-1/2}(\sum_{i=1}^{n} x_i^2 - s_n) = (\sum_{i=1}^{n} x_i)^2/n^{3/2}$ tends to 0 with increasing $n$. Since the convergence of the second moment to 0 implies stochastic convergence to 0, we have:

$$\lim_{k\to\infty}\lim_{n\to\infty} [n^{1/2}(T_{nk}-\mu_k),\ n^{-1/2}(\sum_{i=1}^{n} x_i^2 - n)]$$

$$= \lim_{k\to\infty}\lim_{n\to\infty} [n^{1/2}(T_n-\mu),\ n^{1/2}((s_n/n)-1)]$$

$$= \lim_{n\to\infty} [n^{1/2}(T_n-\mu),\ n^{1/2}((s_n/n)-1)]$$

from which the assertion of the theorem follows by the asymptotic theory of large samples (see Rao [14, Sect. 6a]) and straigthforward calculation.

Remark. If $\sigma > 0$ the theorem gives the proper limiting distribution of $W_n$. In the opposite case the proper limiting distribution - if exists - may be different from normal, as is, for example, the asymptotic distribution of the DeWet-Venter statistic [5].

## 3. Applications

3.1. D'Agostino test. Let be

$$\lim_{n\to\infty} J_n(x)=J(x)=x-(1/2)$$

Consider the following two hypotheses:

$$H_o: f(x) = \varphi(x) = (2\pi)^{-1/2}\, e^{-x^2/2}$$

$H_1: f(x) = f_1(x)$ where $f_1(x) = 1+a-2ax$

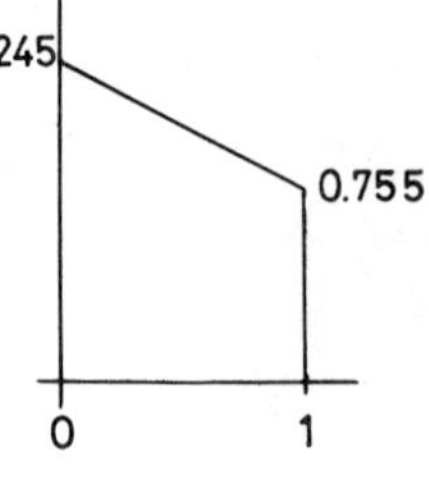


for $0<x<1$ and $f_1(x)=0$ otherwise.
Here $a$ is the positive root of the equation

$$\frac{14700 + 2940a - 11025a^2}{(70+7a-32a^2)^2} = \pi$$

$(a \cong 0.529)$

Theorem 1 is applicable and $\mu$ has under $H_1$ the same value as under $H_o$. The corresponding values of $\sigma^2$ are positive in both cases.

This example shows that the test of normality of D'Agostino [3,4] is inconsistent (the power does not tend to 1 with increasing $n$ at $H_1$). It is not advisable to use this test if general protection is required against all possible alternatives.

This test is contained in the international standard draft ISO/DP 5479 which has been accepted by a majority vote in the technical committee TC 69 of the International Organization for Standardization. The Hungarian Office for Standardization has voted against the draft.

3.2. Shapiro-Wilk test for the exponential distribution. Greenwood test. Consider the following statistic (Shapiro and Wilk [20]):

$$W_n = \frac{n(\bar{X}-X_{1n})^2}{(n-1)\sum_{i=1}^{n}(X_i-\bar{X})^2} \tag{3.1}$$

In this case, the conditions of Theorem 1. concerning $W_n$ are not satisfied. However, the following theorem can be proved:

Theorem 2. Let the distribution function $F(x)$ have the first two moments $a, 1$ and the third and fourth central moments $\sqrt{\beta_1}$ and $\beta_2$, respectively $(a>0)$. Let us suppose, further, that $F(0)=0$ and $F(x)>0$ for $x>0$.

Then

$$\lim_{n\to\infty} \mathcal{L}[n^{1/2}(nW_n - a^2)] =$$

$$\mathcal{N}(0,\ a^4(\beta_2 - 1) - 4a^3\sqrt{\beta_1} + 4a^2)$$

where $W_n$ is given by formula (3.1).

Proof. The straightforward application of the central limit theorem and the theory of large samples leads to the relation:

$$\lim_{n\to\infty} \mathcal{L}(U_n) = \mathcal{N}\{0, [a^2(\beta_2 - 1) - 4a\sqrt{\beta_1} + 4]/4\}$$

where

$$U_n = n^{1/2}\{n^{1/2}\bar{X}[(n-1)\sum_{i=1}^{n}(X_i - \bar{X})^2]^{-1/2} - a\}$$

It is known that $\sum_{i=1}^{n}(X_i - \bar{X})^2/n$ tends stochastically to 1 (variance of $X_1$) and thus, by virtue of the condition, the stochastic limit of

$$V_n = n X_{1n}/[(n-1)\sum_{i=1}^{n}(X_i - \bar{X})^2]^{1/2}$$

is 0. It follows that the limiting distribution of $U_n - V_n$ agrees with that of $U_n$. Using again the theory of large samples we obtain the assertion of the theorem.

A similar theorem can be proved easily for the closely related test of Greenwood [9] which has the test statistic

$$G_n = \frac{\sum_{i=1}^{n} X_i^2}{(\sum_{i=1}^{n} X_i)^2}$$

A special case was obtained by Moran [12] earlier.

Now we consider the following two hypotheses:

$H_0 : f(x)=f_0(x)$ where $f_0(x)=e^{-x}$ if $x>0$

and $f_1(x)=0$ otherwise,

$H_1 : f(x)=f_1(x)$ where $f_1(x)=c\sqrt{(1/x)-1}$ if $0<x<1$

and $f_1(x)=0$ otherwise (c is a norming factor).

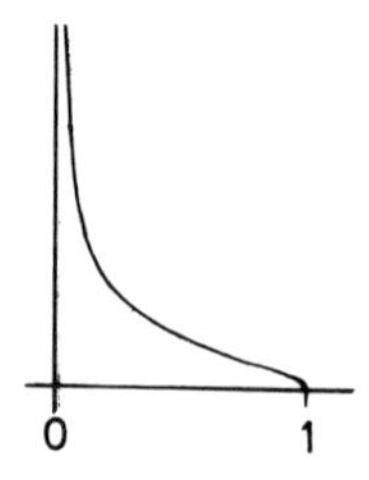


We obtain - just as in the former case - that the test of Shapiro and Wilk for exponentiality is inconsistent and the same is true for Greenwood's test. Moreover, these tests have high asymptotic bias in favor of the considered alternative. For example, if $\alpha=5\%$ is the size of the test, the asymptotic power at the point $f(x)=f_1(x)$ is $0.55\%$.

Earlier critical remarks concerning Greenwood's test are to be found in the discussion of the paper of Greenwood [9] and in Bartholomew [1].

The tests proposed by Hahn and Shapiro [11, § 8.2, Eq. 8.29] and by Stephens [24] are equivalent with Greenwood's test.

4. Acknowledgement.

The author expresses his gratitude to W. Sendler for his helpful comments.

References

[1] Bartholomew, D.J., Testing for departure from the exponential distribution. Biometrika 44(1957) 253-256.

[2] Chernoff, H., Gastwirth, J.L. and Johns, M.V., Asymptotic

distribution of linear combinations of functions of order statistics. Ann.Math.Statist. 38(1967) 52-72.

[3] D'Agostino, R.B., An omnibus test for normality for moderate and large samples. Biometrika 58(1971) 341-348.

[4] D'Agostino, R.B., Small sample probability points for the $D$ -test of normality. Biometrika 59(1972) 219-221.

[5] De Wet, T. and Venter, J.H., Asymptotic distributions of certain test criteria for normality. South Afr. Statist. J. 6(1972) 135-149.

[6] Dyer, A.R., Comparison of tests for normality with a cautionary note. Biometrika 61(1974) 185-189.

[7] Filliben, J.L., The probability plot correlation coefficient test for normality. Technometrics 17(1975) 111-117.

[8] Gerlach, B., A consistent correlation-type goodness-of-fit test. Math. Operationsforsch. Statist., Ser. Statist. 10 /1979/ 427-452.

[9] Greenwood, M., The statistical study of infectious diseases. J.Roy. Statist. Soc. Ser. A. 109(1946) 85-110.

[10] Hahn, G.J. and Shapiro, S.S., Statistical models in engineering. Wiley, New York, 1967.

[11] La Brecque, J., Goodness-of-fit tests based on nonlinearity in probability plots. Technometrics 19(1977) 293-306.

[12] Moran, P.A.P., The random division of an interval.I. J.Roy. Statist. Soc.Suppl. 9(1947) 92-98.

[13] Pearson, E.S., D'Agostino, R.B. and Bowman, K.O., Tests for departure from normality: comparison of powers. Biometrika 64 (1977) 231-246.

[14] Rao, C.R., Linear statistical inference and its applications. 2.ed. Wiley, New York, 1973.

[15] Sarkadi, K., The consistency of the Shapiro-Francia test. Biometrika 62 (1975) 445-450.

[16] Sarkadi, K., Characterization and testing for normality. Colloquia Math. Soc.Bolyai 21, 317-329.

[17] Sarkadi, K., On the consistency of some goodness of fit tests. Proc. 6th Conference Prob. Theory Brasov, Edit.Acad. Romania, Bucuresti (1980).

[18] Shapiro, S.S. and Francia, R.S., Approximate analysis of variance test for normality. J.Amer. Statist. Assoc. 67(1972) 215-225.

[19] Shapiro, S.S. and Wilk, M.B., An analysis of variance test for normality. Biometrika 42(1965) 591-611.

[20] Shapiro, S.S. and Wilk, M.B., An analysis of variance test for the exponential distribution. Technometrics 14(1972) 355-370.

[21] Shapiro, S.S., Wilk, M.B. and Chen,H.J., A comparative study of various tests for normality. J.Amer. Statist. Assoc. 63(1968) 1343-1372.

[22] Shorack, G.R., Functions of order statistics, Ann.Math.Statist. 43(1972)412-427.

[23] Smith, R.B. and Bain, L.J., Correlation type goodness of fit statistics with censored sampling. Comm. Statist. A5(1976) 119-132.

[24] Stephens, M.A., On the W test for exponentiality with origin known. Technometrics 20(1978) 33-35.

# MARTINGALES WITH DIRECTED INDEX SET

Ferenc Schipp

Eötvös Loránd University Budapest

In this paper a theorem of D.L.Burkholder [1] on the square functions is proved for the case when the index set on the martingale is a directed set. Using a generalization of the notion of martingale and stopping time the proof in the predictable case is similar to the original one.

## 1. Introduction

Let $(X,A,P)$ be a probability space, $(A_n, n \in \mathbb{N})$ $(\mathbb{N}:=\{0,1,2,...\})$ a non-decreasing sequence of sub-$\sigma$-fields of $A$. Denote the $L^p(X,A,P)$-norm of the function $f \in L^p := L^p(X,A,P)$ by $\|f\|_p$ $(1 \le p \le \infty)$ and the conditional expectation of $f$ relative to $A_n$ by $E(f|A_n)$. For a martingale $g = (g_n, n \in \mathbb{N})$ relative to $(A_n, n \in \mathbb{N})$ denote the difference sequence of $g$ by $d_n := g_{n+1} - g_n$ $(n \in \mathbb{N})$, the square function of $g$ by

$$Q(g) := \left( \sum_{n=0}^{\infty} |d_n|^2 \right)^{1/2}$$

and the maximal function of $g$ by $g^* := \sup_n |g_n|$. Let furthermore $g_0 = S_0(g_0) = 0$, $Q_n(g) = \left( \sum_{k<n} |d_k|^2 \right)^{1/2}$, $g_n^* = \sup_{1 \le k \le n} |g_k|$ and denote the $L^p$-norm of $g$ by $\|g\|_p := \sup_n \|g_n\|_p$ $(0<p\le\infty)$.

By the well-known martingal maximal inequality (see e.g. [1],[2]) for every $y>0$ and $n \in \mathbb{N}$

(1) $$y\,P\{g_n^*>y\} \le \int_{\{g_n>y\}} |g_n|\,dP \le \|g\|_1 \ .$$

Doob's inequality (see [1])

(2) $$\|g\|_p \le \|g^*\|_p \le \frac{p}{p-1}\|g_p\| \qquad (1<p<\infty)$$

follows immediately from (1).

The following inequalities are due to D.L.Burkholder [1] ; see also [2] and [3].

THEOREMA. *Let* $g = (g_n, n\in\mathbb{N})$ *a martingale. Then*

(3) $$y\,P\{Q(g) > y\} \le 3\,\|g\|_1 \qquad (y>0),$$

*and for every* $p \in (1,+\infty)$

(4) $$c_p \| Q(g)\|_p \le \|g\|_p \le C_p \|Q(g)\|_p \ ,$$

*where* $C_p = O(pq^{1/2})$ , $c_p^{-1} = O(p^{1/2}q)$ *and* $1/p + 1/q = 1$ .

In this paper we generalize the inequality (3) for some martingales with directed index set. We will assume that the countable ordered index set $(N,\le)$ has the following properties:

a/ for every two elements $m,n \in N$ there exists the upper envelope $m \vee n := \sup\{m,n\}$ ,

b/ for every $m \in N$ the set $\{n\in N : n\le m\}$ is finite,

c/ for all $m \in N$ the set $\{n\in N : m\le n\}$ is linearly ordered.

From these conditions it follows easily, that every non-empty subset of $N$ has a minimal element. Furthermore, let min H denote the set of minimum elements of H, $M := \{\min H : H \subseteq N\}$

and $\tilde{N} := \{(n,m) \in N \times N : n \le m \text{ or } m \le n\}$ denote the set of comparable elements of $N^2$. We introduce a relation in the set $M$, denoted by $\le$ : For $A,B \in M$ let $A \le B$ if for all $b \in B$ there exists an element $a \in A$, for which $a \le b$. It is easy to see, that the newly defined relation is a partial ordering in the set $M$, for which $M$ is a net and $A \vee B := \sup\{A,B\} = \min\{a \vee b : a \in A, b \in B\}$, $A \wedge B := \inf\{A,B\} = \min A \cup B$.

We introduce the following generalization of the notion of *stopping time*. Let us fix an increasing sequence of $\sigma$-fields $A_n \subseteq A$ $(n \in N)$. We say, that *the mapping* $\tau : X \to M$ *is a stopping time with respect to the sequence* $(A_n, n \in N)$, *if for all* $n \in N$ $\{n \in \tau\} := \{x \in X : n \in \tau(x)\} \in A_n$ *holds*. It is obvious that in the case $N = \mathbb{N}$ $\tau$ is a stopping time in the usual sense. The constant functions $\bar{n} : X \to M$ defined by $\bar{n}(x) := \{n\}$ $(x \in X, n \in N)$ are obvious stopping times. It can be easily proved, that the mapping $\tau : X \to M$ is a stopping time for $(A_n, n \in M)$ if and only if for every element $n \in N$ $\{\tau \le \bar{n}\} :=$ $= \{x \in X : \tau(x) \le \bar{n}(x)\} \in A_n$ holds. Let $T$ denote the set of stopping times with respect to the sequence $(A_n, n \in N)$. It is easy to prove, that for all elements $\tau, \sigma \in T$ $\tau \vee \sigma, \tau \wedge \sigma \in T$ and consequently $T$ is a net.

In the following we fix an increasing sequence $B_n \subseteq A$ $(n \in N)$ of $\sigma$-fields, and suppose, that $B_k = B_\ell$ if $k_+ = \ell_+$. Here $k_+ := \inf\{s \in N : k < s\}$ and $k < s$ denotes that $k \le s$ and $k \ne s$. Let us introduce the increasing sequence of $\sigma$-fields $A_n := B_m$ $(m_+ = n)$, $A_n := B_n$ $(n_+ = m_+,\ n \in N_o := \min N)$. We investigate a sequence $(E_n, n \in N)$ of operators with the following properties:

a/ $E_n : L^1 \to L^1$ $(n \in N)$ is $B_n$-linear, i.e. for $\alpha, \beta \in L^\infty(B_n) := L^\infty(X, B_n, P)$ and $f,g \in L^1$ we have $E_n(\alpha f + \beta g) = \alpha E_n f + \beta E_n g$ ;

(5)

b/ $E_n \circ E_m = E_{n \wedge m} \quad ((n,m) \in \tilde{N})$ , $E_n \circ E_m = 0 \quad ((n,m) \in N^* := N^2 \setminus \tilde{N})$;

c/ $|E_n f|$ is $\mathcal{B}_n$-measurable $(f \in L^1)$ ;

d/ $E_n : L^2 \to L^2$ is self-adjoint $(n \in N)$ ;

e/ $\|E_n f\|_p \le \|f\|_p \quad (1 \le p \le \infty,\ n \in N)$ .

In the case $N = \mathbb{N}$ the conditions (5) are satisfied for the operators $E_n f := E(f \mid \mathcal{B}_n) \quad (f \in L^1,\ n \in N)$, where $E(f \mid \mathcal{B}_n)$ denotes the conditional expectation operator with respect to the $\sigma$-field $\mathcal{B}_n$.

There comes another important example.Let $N := \{[k2^n,(k+1)2^n) : k,n \in \mathbb{N}\}$
$E_A := \sum_{n \in A} P_n$ , $\mathcal{B}_A := \sigma\{\Psi_1, \Psi_2, \ldots, \Psi_{2^n-1}\}$ $(A = [k2^n,(k+1)2^n) \in N)$,
where $(\Psi_n, n \in \mathbb{N})$ denotes the orthonormed Walsh-Paley system and $P_n f := (\int_0^1 f\Psi_n\,dx)\Psi_n \quad (n \in \mathbb{N})$ are the Walsh-projections. Then for the set $N$ with the inclusion, the mentioned conditions are satisfied and by $E_A f = \Psi_{k2^n} E(f\Psi_{k2^n} \mid \mathcal{B}_A) \quad (A = [k2^n,(k+1)2^n \in N)$ (5) also holds

We introduce the following notions. The sequence $f = (f_n, n \in N)$ of the functions $f_n \in L^1 \ (n \in N)$ *is called an orthogonal martingale* (briefly OM) with respect to $(E_n, n \in N)$, if $f_n = E_n f_m$ for every $n,m \in N$, $n \le m$. If $N = \mathbb{N}$ and $E_n h = E(h \mid \mathcal{B}_n) \ (n \in N)$ , then $f$ is a martingale in the usual sense.

For an OM denote $d_n := f_{n+} - f_n$ , $f_n^* := \sup_{m \le n} |f_m| \ (n \in N)$ ,
$f_n^- = \sup_{m_+ = n} |f_m| \ (n \in N \setminus N_o)$ , and $f_n^- = \sup_{m_+ = n_+} |f_m| \ (n \in N_o)$. Further, let

$$Q_n f := \sup\{(\sum_{m \le k < n} |d_k|^2)^{1/2} : m \le n,\ m \in N_o\} \quad (n \in N \setminus N_o),$$

$$Q_n f = 0 \quad (n \in N_o) .$$

We say, that the OM $f = (f_n, n \in N)$ *is predictable* if there exists a constant $c > 0$ such that

$$|f_n| \leq c\, f_n^- \qquad (n \in N) . \tag{6}$$

We can prove the following generalization of the martingale maximal inequality (see [4],[5]).

THEOREM B. *Let* $f = (f_n, n \in N)$ *be an OM. Then*

$$\tau_y(x) := \min\,\{n \in N;\ f_n^-(x) > y\} \qquad (y > 0,\ x \in X) \tag{7}$$

*is a stopping time with respect to* $(A_n, n \in N)$ *and*

$$y^2 \sum_{m<n} P\{n \in \tau_y\} \leq \int_{\{f_n^* > y\}} |f|^2\, dP \quad (y > 0, n \in N). \tag{8}$$

On the basis of (8) it is obvious, that

$$y^2\, P\{f_n^* > y\} \leq \int_{\{f_{n_+}^* > y\}} |f|^2\, dP \tag{8'}$$

and in the case $N = \mathbb{N}$ (8) and (8′) are equivalent.

From this we obtain an analogue of the Doob-inequality

$$\|f^*\|_p \leq \left(\frac{p}{p-2}\right)^{1/2} \|f\|_p \qquad (2 < p < \infty) . \tag{8''}$$

In this paper we give a generalization of the inequality (3).

THEOREM *Let* f *be a predictable OM. Then*

$$y^p\, P\{Q^* f > y\} \leq K(p,c)\, \|f\|_p \qquad (y > 0, 2 \leq p < \infty), \tag{9}$$

*where* $Q^* f = \sup_{n \in N} Q_n f$ *and* $K(p,c)$ *depends only on* p *and on the constant* c *in* (6).

## 2. Proof of the Theorem

Starting from a spectral decomposition, we give an useable estimation for the sublinear operator $Q^*$. Let us fix an $N \in N$ index and an increasing sequence $\lambda = (\lambda_m, m \in N)$, consisting of non--negative functions, for which $\lambda_m$ is $A_m$-measurable and

$$|E_m f| \leq \lambda_m \qquad (m \in N) .$$

By the condition (6) $\lambda_m := \sup_{k \leq m} (cf_k^-)$ $(m \in N)$ is such a sequence. We denote the indicator function of the set $H$ by $I(H)$. It is easy to see, that by the sequence

$$\text{(10)} \qquad \varepsilon_m(y) := 2I\{\tfrac{1}{2}\lambda_{m_+} \leq y < \lambda_{m_+}\}/(\lambda_{m_+} \vee y) \qquad (y > 0, m \in N)$$

the operator $S_n f := (\sum_{n \leq m \leq N} |d_m|^2)^{1/2}$ $(f = (f_m, m \in N)$ can be written in the form

$$\text{(11)} \qquad S_n f = (\sum_{n \leq m \leq N} (d_m \int_0^{+\infty} \varepsilon_m(y)\, dy)^2)^{1/2}$$

furthermore, for the function $F_n^y := \sum_{n \leq m \leq N} \varepsilon_m(y)\, d_m$ by (4) we have

$$\text{(12)} \qquad \begin{aligned} &|F_n^y| \leq 8I\{y \leq \lambda^*\} && (y > 0) , \\ &\| I(H)\, S_n^\varepsilon f \|_p \leq R_p \| I(H) \|_p && (1 < p < \infty) \end{aligned}$$

where $S_n^\varepsilon f := (\sum_{n \leq m \leq N} |\varepsilon_m(y) d_m|^2)^{1/2}$, $\lambda^* := \sup_{m \in N} \lambda_m$, $H \in A_n$ and $R_p$ depends only on $p$. The stopping time

$$\text{(13)} \qquad \sigma_y(x) := \min \{m \in N : \lambda_m(x) > y\} \qquad (x \in X, y > 0)$$

is closely connected with the sequence $(\varepsilon_m(y), m \in N)$ and with $\tau_y$, namely $\sigma_y = \tau_{y/c}$ and $\{\lambda_m > y\} = \{\sigma_y \leq \bar{m}\}$ $(m \in N, y > 0)$.

From (10) and (11) it follows, that

$$S_n f \le \int_0^{+\infty} \Big( \sum_{n\le m\le N} |\varepsilon_m(y) d_m|^2 \Big)^{1/2} dy \le$$

$$\le \int_0^{+\infty} \Big( \sum_{\bar{n}\vee\sigma_y \le m_+ < (\bar{n}\wedge\bar{N}\vee\sigma_{2y})} |\varepsilon_m(y) d_m|^2 \Big)^{1/2} dy \le$$

$$\le 2\lambda^* + \int_0^{+\infty} \sup_{r\in N} I\{r\in\sigma_y\} S_r^\varepsilon f \, dy ,$$

and consequently the following estimations hold:

$$\Omega_N^* f := \sup\{ \Big( \sum_{n\le m\le N} |d_m|^2 \Big)^{1/2} : n \le N, n\in N_o \} \le$$

$$\le 2\lambda^* + \int_0^{+\infty} T_N^y f \, dy \quad ,$$

where $T_N^y f := \sup_{r\in N} I\{r\in\tau_y\} S_r^\varepsilon f$ , and similarly

$$I\{\lambda^* \le y_o\} \Omega_N^* f \le 2y_o + \int_0^{y_o} T_N^y f \, dy \quad . \tag{14}$$

By (12),(13) and (8).

$$\| \int_0^{y_o} T_N^y f \, dy \|_{2p} \le \int_0^{y_o} \Big( \sum_{r\in N} \| I\{r\in\sigma_y\} S_r^\varepsilon f \|_{2p}^{2p} \Big)^{1/2p} dy \le$$

$$\le R_{2p} \int_0^{y_o} \Big( (y/c)^{-2} \int_{\{f_N^* > y/c\}} |f_N|^2 \, dP \Big)^{1/2p} dy \le$$

$$\le R_{2p}\, c^{1/2} \Big( \int_0^{y_o} y^{-1/2} \, dy \Big) \Big( \int_X |f_N^*|^p \, dP \Big)^{1/2p} ,$$

and by (14) we get

$$y_o (P\{\Omega_N^* f > 3y_o\})^{1/p} \le \tilde{K}(p,c) \, \|f\|_p \qquad (y_o > 0) ,$$

where $\tilde{K}(p,c)$ depends only on $p$ and $c$. From this taking the limit as $N\to\infty$ and using (8″) we obtain (9).

REFERENCES

[1] BURKHOLDER,D.L., Distribution function inequalities for martingales, *Annals of Prob.* 1(1973), 19-42.

[2] GARSIA,A., Martingale Inequalities, *Seminar Notes on Recent Progress, W.A.Benjamin, Reading, Massachusetts* 1973.

[3] NEVEU,J., Discrete-parameter martingales, *North-Holland Math. Library, Amsterdam, Oxford, New-York,* 1975.

[4] SCHIPP,F., Fourier series and martingale transforms. Linear Spaces and Approximations. Ed. P.L.Butzer and B.Sz.-Nagy. *Birkhäuser Verlag Basel,* 1978, 571-581.

[5] SCHIPP,F., On Carleson's method. *Coll. Math. Soc. J.Bolyai,* 19; *Fourier Analysis and Approximation Theory, Budapest* (1976); 679-695 .

# EXTENSIONS OF PARTIAL HOMOMORPHISMS IN PROBABILITY THEORY

Gábor J. Székely

Eötvös University, Budapest

Two years ago in Amsterdam when trying to solve a problem about invariant measures on semigroups A.A.Balkema and myself arrived to the following very simple question:
Are there continuous nontrivial homomorphisms from the convolution semigroup of probability distribution functions $\mathcal{F}$ /the topology is the Levy metric/ to the usual additive topological group of the real line $\mathcal{R}$. /A homomorphism is trivial if the image of every element is 0./ At that time we could not find the answer to this question. One year later G.Halász proved that the answer is negative. In fact a stronger result is also true: there is no continuous homomorphism from $\mathcal{F}$ to the complex unit circle /the operation is the complex multiplication and the topology is the usual one/ except the trivial homomorphism. In the theory of topological groups these type of continuous homomorphisms /the group representations/ play very important role. The above mentioned result states that there is no continuous representation of the semigroup $\mathcal{F}$. A natural further question is the following:
Are there /non-continuous, nontrivial/ homomorphisms from the semigroup $\mathcal{F}$ to the group $\mathcal{R}$. Here the answer is affirmative. In [6] we have proved somewhat more: There exists a homomorphism $\varphi$ from $\mathcal{F}$ to $\mathcal{R}$ such that $\varphi(F) = E(F)$ for every $F \in \mathcal{F}$ having finite expectation $E(F)$. This theorem was implied by the following algebraic
<u>LEMMA</u> /for the proof see [6]/.
Let $S$ be a commutative semigroup, $S'$ a subsemigroup of $S$, $G$ a

group and $\psi:S' \to G$ a homomorphism. For the existence of a homomorphism $\varphi:S \to G$ satisfying $\varphi(s) = \psi(s)$ for every $s \in S'$ the following condition is necessary:

$$ss_1 = ss_2 \qquad \text{implies} \quad \psi(s_1) = \psi(s_2) \qquad /1/$$
$$\text{if} \quad s \in S \, , \quad s_1, s_2 \in S'$$

If $G$ is divsible then /1/ is also sufficient.

Now we apply our lemma for renewal sequences. Let $X$ be a non-negative integer-valued random variable, and let $X_1, X_2, \ldots$ be independent copies of it. Denote

$$S_n = X_1+X_2+\ldots+X_n \qquad n=1,2,\ldots$$

As it is well-known the sequence of real numbers $u=(u_1,u_2,\ldots)$ is called a renewal sequence if there exists an $X$ such that $u_n$ = pr $/S_k = n$ for some $k/$ $n=1,2,\ldots$ . An interesting property of these renewal sequences is that they from a semigroup if the operation is the componentwise product /[5]/. This observation was the starting point of many interesting research about the arithmetic properties of these sequences and their generalisations /[3],[4]/. Without the essential loss of generality it is enough to consider only the so-called aperiodic renewal sequences defined by the condition g.c.d. $\{n : u_n > 0\} = 1$ . The semigroup of aperiodic renewal sequences will be denoted by $\mathcal{U}$.

An interesting theorem of R.Davidson [1] states that if $\mathcal{U}$ is endowed with the topology of term-by-term convergence then there is no continuous homomorphism from $\mathcal{U}$ to $\mathcal{R}$ /except the trivial one/. The problem, whether there is at all /non-continuous/ homomorphism from $\mathcal{U}$ to $\mathcal{R}$ still remained open. Now we show the existence of such a homomorphism.

By the Erdős-Feller-Pollard theorem /[2]/ for every $u \in \mathcal{U}$ $\lim_{n\to\infty} u_n$ always exists and it is positive if and only if $E(X) < \infty$. Let $S'$ be the following subsemigroup of $\mathcal{U}$ : $u \in S'$ if and only if $\lim_{n\to\infty} u_n > 0$. Let $\psi(u) = \log \lim u_n$ if $u \in S'$. This is evidently a homomorphism from $S'$ to $\mathcal{R}$. For the proof of the property /1/ it is enough to observe that every renewal sequence has infinitely many non-zero component $u_n$ thus if $u_n u_{n1} = u_n u_{n2}$ for $n=1,2,\ldots$ where $(u_n) \in \mathcal{U}$, $(u_{n1})$ and $(u_{n2}) \in S'$ then $\lim_{n\to\infty} u_{n1} = \lim_{n\to\infty} u_{n2}$. By our Lemma we obtained the following

THEOREM. There exists a homomorphism $\varphi$ from $\mathcal{U}$ to $\mathcal{R}$ such that

$$\varphi(u) = \log \lim_{n\to\infty} u_n \qquad \text{whenever} \quad u \in S'$$

DEFINITION. Let $S$ be an arbitrary semigroup. The dimension of $S$ with respect to $\mathcal{R}$ is infinite if for every finite subset of $S$ there is at least one homomorphism $\gamma$ from $S$ to $\mathcal{R}$ such that the $\gamma$-image of this finite subset is $0 \in \mathcal{R}$ but $\gamma(S) \neq 0$. /A general definition of algebraic dimension of this type can be found in [7]./

AN OPEN PROBLEM. Is the dimension of $\mathcal{F}$ and $\mathcal{U}$ infinite?

## REFERENCES


[1] DAVIDSON,R.: *Arithmetic and other properties of certain Delphic semigroups.* I.Ztschr. Wahrsch'theorie verw. Geb. 10/1968/, 120-145.

[2] ERDŐS,P.;FELLER,W.; POLLARD,H.: *A theorem on power series.* Bull. Amer. Math. Soc. 55/1946/, 292-302.

[3] KENDALL,D.G.: *Renewal sequences and their arithmetic.* Symposium on Probability Methods in Analysis /Lecture Notes in Math. 31/ Springer, Berlin /1967/.

[4] KENDALL,D.G.: *Delphic semigroups, infinitely divisible regenarative pheonomena, and the arithmetic of p-functions.* Ztschr. Wahrsch' theorie verw. Geb. 9/1968/, 163-195.

[5] KINGMAN,J.F.C.: *Regenerative phenomena.* Wiley, London /1972/.

[6] RUZSA,I.Z., SZÉKELY,G.J.: *An extension of expectation.* /submitted to the Ztschr.Wahrsch'theorie verw.Geb./

[7] SZÉKELY,G.J.: *Algebraic dimension of semigroups with application to invariant measures.* Semigroup Forum 17/1979/ 185-187.

# A Remark on the Strong Law of Large Numbers for Random Indexed Sums

Dominik Szynal

M. Curie-Skłodowska Uniwersity , Lublin

Summary. The aim of this note is to give estimates for the mathematical expectation of the so-called counting random variable appearing in the strong law of large numbers.

1. Let $\{X_k, k \geq 1\}$ be a sequence of independent random variables with $EX_k = 0$, $k \geq 1$, and let $\{\nu_k, k \geq 1\}$ be a sequence of positive integer-valued random variables defined on the same probability space $(\Omega, \mathcal{A}, P)$ as $\{X_k, k \geq 1\}$. Denote by $\nu$ a random variable taking values from an interval $(c,d)$, $0 < c < d < \infty$

Put for any given $\varepsilon > 0$

$$N_n^+(\varepsilon) = \sum_{k=1}^n I\left[\,\left|\frac{\nu_k}{k} - \nu\right| \geq \varepsilon\right]$$

and

$$N_n(\varepsilon) = \sum_{k=1}^n I\left[\,\left|\frac{S_{\nu_k}}{\nu_k}\right| \geq \varepsilon\right] ,$$

where $I[A]$ denotes the indicator of the event $[A]$ and

$$S_m = \sum_{k=1}^m X_k$$

We are interested in finding estimates on $EN_n(\varepsilon)$ in terms of $EN_n^+(\varepsilon)$ and some characteristics of the random variables $X_k$ for diffrent classes of random variables.

We start with the following two lemmas.

Lemma 1 [2]. Let $X_1, X_2, \ldots, X_n$ be independent random variables such that

$X_i \leq L_i$, $EX_i = 0$, $i = 1,2,\ldots,n$, where $L_i$, $i = 1,2,\ldots,n$ are positive constants.

Then

$$(1)\quad P\left[S_n \geqslant x\right] \leqslant \exp\left\{\frac{x}{L} - \left(\frac{x}{L} + \frac{s_n^2}{L^2}\right)\ln\left(\frac{xL}{s_n^2}+1\right)\right\},$$

where $S_n = \sum_{i=1}^{n} X_i$, $L = \max\left\{L_1, L_2, \ldots, L_n\right\}$, $s_n^2 = \sum_{k=1}^{n} \sigma_k^2$.

Lemma 2 [1]. Under the assumption of Lemma 1

$$(2)\quad P\left[\max_{k\leqslant n} S_k \geqslant x\right] \leqslant \exp\left\{\frac{x}{L} - \left(\frac{x}{L} + \frac{s_n^2}{L^2}\right)\ln\left(\frac{xL}{s_n^2}+1\right)\right\}$$

2. Let us first consider estimates on $EN_n(\varepsilon)$ in the case when $X_1, X_2, \ldots, X_n$ are independent identically distributed random variables.

Lemma 3. Let $\left\{X_k, k \geqslant 1\right\}$ be a sequence of independent identically distributed random variables with $EX_1 = 0$, and let $\left\{\nu_k, k \geqslant 1\right\}$ be a sequence of positive integer-valued random variables defined on the same probability space $(\Omega, \mathcal{A}, P)$ as $\left\{X_k, k \geqslant 1\right\}$. Denote by $\nu$ a random variable taking values from an interval $(c,d)$, $0 < c < d < \infty$.

Then for any given $\varepsilon > 0$ and a fixed $\delta \in (0, \frac{1}{3})$

$$(3)\quad EN_n(\varepsilon) = O\Big(EN_n^+(\varepsilon) + \sum_{k=1}^{n} k\, P\left[|X_1| \geqslant \tfrac{1}{8}(c-\varepsilon)(1-3\delta)\varepsilon k\right] + \sum_{k=1}^{n}\left(EX_1^2 I\left[|X_1| < \tfrac{1}{8}(c-\varepsilon)(1-3\delta)\varepsilon k\right]/k\right)^{1+\frac{\delta}{1-2\delta}}\Big)$$

as $n \to \infty$.

Proof. It is not difficult to see that

$$(4)\quad EN_n(\varepsilon) = \sum_{k=1}^{n} P\left[\left|\frac{S_{\nu_k}}{\nu_k}\right| \geqslant \varepsilon\right] \leqslant$$

$$\leqslant \sum_{k=1}^{n} P\left[\left|\frac{\nu_k}{k} - \nu\right| \geqslant \varepsilon\right] + \sum_{k=1}^{n} P\left[\left|\frac{S_{\nu_k}}{\nu_k}\right| \geqslant \varepsilon, \left|\frac{\nu_k}{k} - \nu\right| < \varepsilon\right] \leqslant$$

$$\leqslant \sum_{k=1}^{n} P\left[\left|\frac{\nu_k}{k} - \nu\right| \geqslant \varepsilon\right] + \sum_{k=1}^{n} P\left[\left|\frac{S_{\nu_k}}{\nu_k}\right| \geqslant \varepsilon, a(k) \leqslant \nu_k \leqslant b(k)\right],$$

where $a(k) = [(c-\varepsilon)k]$, $b(k) = [(d+\varepsilon)k] + 1$.

But

$$P\left[|S_{\nu_k}| \geqslant \nu_k \varepsilon, a(k) < \nu_k < b(k)\right] \leqslant$$

$$\leq P\left[|S_{\nu_k}| \geq a(k)\varepsilon,\ a(k) < \nu_k < b(k)\right] \leq$$

$$\leq P\left[|S_{a(k)}| \geq a(k)\varepsilon/2\right] + P\left[\max_{a(k)\leq l\leq b(k)} |S_l - S_{a(k)}| \geq a(k)\varepsilon/2\right]$$

Define for $1 \leq i \leq k$ and $\delta > 0$

$$X_i^* = \begin{cases} X_i & \text{if} \quad |X_i| \leq L_k , \\ 0 & \qquad |X_i| > L_k , \end{cases}$$

where

$$L_k = \frac{1}{8}(c-\varepsilon)(1-3\delta)\varepsilon k$$

and put $\quad S_k^* = \sum_{i=1}^{k} X_i^*$

Let $E_k$ stand for the event

$$E_k = \left[\max_{a(k)\leq l\leq b(k)} |S_l - S_{a(k)}| \neq \max_{a(k)\leq l\leq b(k)} |S_l^* - S_{a(k)}^*|\right]$$

Then we have

$$(5) \quad P\left[|S_{a(k)}| \geq a(k)\varepsilon/2\right] \leq P\left[|S_{a(k)}^*| \geq \frac{a(k)\varepsilon}{4}\right] + \sum_{l=1}^{a(k)} P\left[|X_l| \geq L_k\right]$$

and

$$(6) \quad P\left[\max_{a(k)\leq l\leq b(k)} |S_l - S_{a(k)}| \geq \frac{a(k)\varepsilon}{2}\right] \leq$$

$$\leq P\left[E_k\right] + P\left[\max_{a(k)\leq l\leq b(k)} |S_l^* - S_{a(k)}^*| \geq \frac{a(k)\varepsilon}{2}\right].$$

But

$$(7) \quad P\left[E_k\right] = \sum_{l=a(k)}^{b(k)} P\left[|X_l| \geq L_k\right],$$

and

$$(8) \quad P\left[\max_{a(k)\leq l\leq b(k)} |S_l^* - S_{a(k)}^*| \geq \frac{a(k)\dot{\varepsilon}}{2}\right] \leq$$

$$\leq P\left[|S_{a(k)}^*| \geq \frac{a(k)\varepsilon}{4}\right] + P\left[\max_{1\leq l\leq b(k)} |S_l^*| \geq \frac{a(k)\varepsilon}{4}\right].$$

Set $\quad X_i' = X_i^* - EX_i^*, \quad S_k' = \sum_{i=1}^{k} X_i'$ .

Then we have

$$P\left[|S_{a(k)}^*| \geq \frac{a(k)\varepsilon}{4}\right] \leq P\left[|S_{a(k)}'| + |ES_{a(k)}^*| \geq \frac{a(k)\varepsilon}{4}\right] =$$

$$= P\left[ |S'_{a(k)}| + |ES^{**}_{a(k)}| \geq \frac{a(k)\varepsilon}{4} \right],$$

where $X_i^{**} = X_i I[|X_i| \geq L_k]$ , $S_m^{**} = \sum_{i=1}^{m} X_i^{**}$ .

Now taking into account that

$$|ES^{**}_{a(k)}| \leq a(k) E|X_1| I[|X_1| \geq L_k]$$

and that $E|X_1| I[|X_1| \geq L_k] \to 0$ as $k \to \infty$ ,

we obtain for sufficiently large k

$$P\left[ |S^*_{a(k)}| \geq \frac{a(k)\varepsilon}{4} \right] \leq P\left[ |S'_{a(k)}| \geq \frac{a(k)\varepsilon}{4} (1-\delta) \right].$$

But, by the assumption that $EX_i = 0$ , $i \geq 1$ , and the fact $E|X_i^{**}| \to 0$ as $k \to \infty$ we get

$$|X_i'| \leq |X_i^*| + |EX_i^*| = |X_i^*| + |EX_i^{**}| \leq$$

$$\leq L_k + o(1) \leq \frac{1}{8}(c-\varepsilon)(1-2\delta)\varepsilon k = L_k'$$

for sufficiently large k, for $k \geq k_0 = k_0(\delta)$ say.

Taking into account that $[x] > \frac{1}{2}x$ , and using the facts given above and Lemma 1 , we get

$$(9) \quad P\left[ |S^*_{a(k)}| \geq \frac{a(k)\varepsilon}{4} \right] \leq P\left[ |S'_{a(k)}| \geq \frac{a(k)\varepsilon}{4}(1-\delta) \right] \leq$$

$$\leq 2 \exp\left\{ \frac{2(1-\delta)a(k)}{(c-\varepsilon)(1-2\delta)k} - \left( \frac{2(1-\delta)a(k)}{(c-\varepsilon)(1-2\delta)k} + \frac{\sigma^2 S'_{a(k)}}{(L_k')^2} \right) \ln\left( \frac{\varepsilon^2(1-\delta)(1+2\delta)(c-\varepsilon)^2 k^2}{64 \sum_{i=1}^{a(k)} \sigma^2 X_i'} + 1 \right) \right\}$$

$$\leq 2 \exp\left\{ \frac{2(1-\delta)}{1-2\delta} - \frac{1-\delta}{1-2\delta} \ln\left( \frac{\varepsilon^2(1-\delta)(1-2\delta)k^2}{64 \sum_{i=1}^{a(k)} \sigma^2 X_i'} + 1 \right) \right\}$$

$$\leq 2 e^{\frac{2(1-\delta)}{1-2\delta}} \cdot \left( \frac{64 \sum_{i=1}^{a(k)} \sigma^2 X_i'}{\varepsilon^2(1-\delta)(1-2\delta)k^2} \right)^{1+\frac{\delta}{1-2\delta}}$$

$$\leq 2^{7+\frac{6\delta}{1-2\delta}} e^{\frac{2(1-\delta)}{1-2\delta}} \left[ \varepsilon^2(1-\delta)(1-2\delta)(c-\varepsilon)^2 \right]^{-\frac{1-\delta}{1-2\delta}} \left( \frac{\sum_{i=1}^{a(k)} E(X_i^*)^2}{k^2} \right)^{1+\frac{\delta}{1-2\delta}}$$

$$\leq M(c,\delta,\varepsilon) \left( \frac{\sum_{i=1}^{a(k)} E(X_i^*)^2}{k^2} \right)^{1+\frac{\delta}{1-2\delta}} ,$$

where $M(c,\delta,\varepsilon)$ is a positive constant depending only on $c, \delta, \varepsilon$.

We shall now estimate $P\left[\max_{1\le l\le b(k)} |S_l^*| \ge \frac{a(k)\varepsilon}{4}\right]$.

Note that

$$P\left[\max_{1\le l\le b(k)} |S_l^*| \ge \frac{a(k)\varepsilon}{4}\right] \le$$

$$\le P\left[\max_{1\le l\le b(k)} \left(|S_l^* - ES_l^*| + |ES_l^*|\right) \ge \frac{a(k)\varepsilon}{4}\right]$$

$$\le P\left[\max_{1\le l\le b(k)} \left(|S_l'| + \sum_{i=1}^{l} E|X_i|\, I\left[|X_i| \ge L_k\right]\right) \ge \frac{a(k)\varepsilon}{4}\right]$$

$$\le P\left[\max_{1\le l\le b(k)} |S_l'| \ge \frac{a(k)\varepsilon}{4} - b(k)\, E|X_1|\, I\left[|X_1| \ge L_k\right]\right]$$

$$\le P\left[\max_{1\le l\le b(k)} |S_l'| \ge \frac{a(k)\varepsilon}{4}(1-\delta)\right]$$

for sufficiently large k.

Lemma 2 and the evaluations similar to those using in the proof of Lemma 1 allow to write

$$(10)\quad P\left[\max_{1\le l\le b(k)-a(k)} |S_l'| \ge \frac{a(k)\varepsilon}{4}(1-\delta)\right] \le$$

$$\le M(c,\varepsilon,\delta)\left(\frac{\sum_{i=1}^{b(k)} E(X_i^*)^2}{k^2}\right)^{1+\frac{\delta}{1-2\delta}}$$

Putting the inequalities (4)-(10) together and taking $d \ge 1$ ( without loss of generality ), we get

$$(11)\quad \sum_{k=1}^{m} P\left[\left|\frac{S_{\nu_k}}{\nu_k}\right| \ge \varepsilon\right] \le \sum_{k=1}^{m} P\left[\left|\frac{\nu_k}{k} - \nu\right| \ge \varepsilon\right] +$$

$$+ \sum_{k=1}^{m} b(k) P\left[|X_1| \ge L_k\right] + 3M(c,\varepsilon,\delta) \sum_{k=1}^{m} \left(\frac{\sum_{i=1}^{b(k)} E(X_i^*)^2}{k^2}\right)^{1+\frac{\delta}{1-2\delta}} \le$$

$$\le \sum_{k=1}^{m} P\left[\left|\frac{\nu_k}{k} - \nu\right| \ge \varepsilon\right] + 3d \sum_{k=1}^{m} k P\left[|X_1| \ge L_k\right] +$$

$$+ 3M(c,\varepsilon,\delta) \sum_{k=1}^{m} \left(\frac{\sum_{i=1}^{3dk} E(X_i^*)^2}{k^2}\right)^{1+\frac{\delta}{1-2\delta}}$$

Hence , by (11), we have

$$\sum_{k=1}^{n} P\left[\left|\frac{S_{\nu_k}}{\nu_k}\right| \geq \varepsilon\right] \leq \sum_{k=1}^{n} P\left[\left|\frac{\nu_k}{k} - \nu\right| \geq \varepsilon\right]$$

$$+ 3d \sum_{k=1}^{n} k P\left[|X_1| \geq L_k\right] + 9d^2 M(c,\varepsilon,\delta) \sum_{k=k_0}^{n} \left(\frac{E X_1^2 I[|X_1| < L_k]}{k}\right)^{1+\frac{\delta}{1-2\delta}}$$

$$+ \frac{3k_0(k_0+1)}{2},$$

which proves (3).

Lemma 3 allows us to give the following theorem.

Theorem 1. If under the assumptions of Lemma 3, $EX_1^2 < \infty$ and for any given $\varepsilon > 0$ $EN_\infty^+(\varepsilon) < \infty$ then $EN_\infty(\varepsilon) < \infty$. More precisely, for any given $\varepsilon > 0$ and a fixed $\delta \in (0, \frac{1}{3})$

$$\sum_{n=1}^{\infty} P\left[|S_{\nu_n}| \geq \varepsilon \nu_n\right] \leq K_1 \left(\sum_{n=1}^{\infty} P\left[|\nu_n - n\nu| \geq n\varepsilon\right] + \right.$$

$$\left. + \sum_{n=1}^{\infty} n P\left[|X_1| \geq \frac{1}{8}(c-\varepsilon)(1-3\delta)\varepsilon n\right] + \sum_{n=1}^{\infty} \left(EX_1^2 I\left[|X_1| < \frac{1}{8}(c-\varepsilon)(1-3\delta)\varepsilon n\right]/n\right)^{1+\frac{\delta}{1-2\delta}}\right),$$

where $K_1$ is a positive constant.

We now discuss sequences of non-identically distributed random variables.

First we consider a sequence $\{X_n, n \geq 1\}$ of uniformly integrable random variables, i.e. satisfying the condition

$$\lim_{\alpha \to \infty} \sup_k E|X_k| I\left[|X_k| \geq \alpha\right] = 0.$$

Lemma 4. Let $\{X_k, k \geq 1\}$ be a sequence of independent uniformly integrable random variables with $EX_k = 0$, $k \geq 1$, and let $\{\nu_k, k \geq 1\}$ be a sequence of positive integer-valued random variables defined on the same probability space $(\Omega, \mathcal{A}, P)$ as $\{X_k, k \geq 1\}$. Denote by $\nu$ a random variable taking values from an interval $(c,d)$, $0 < c < d < \infty$.

Then for any given $\varepsilon > 0$ and a fixed $\delta \in (0, \frac{1}{3})$

$$(12)\quad EN_m(\varepsilon)=O\Big(EN_m^+(\varepsilon)+\sum_{k=1}^{m}\sum_{i=1}^{3dk}P\big[|X_i|\geq \tfrac{1}{8}(c-\varepsilon)(1-3\delta)\varepsilon k\big]+$$

$$+\sum_{k=1}^{m}\Big(\sum_{i=1}^{3dk}EX_i^2 I\big[|X_i|<\tfrac{1}{8}(c-\varepsilon)(1-3\delta)\varepsilon k\big]/k^2\Big)^{1+\frac{\delta}{1-2\delta}}\Big).$$

Proof. The considerations given in the proof of Lemma 3 give the estimate

$$(13)\quad EN_m(\varepsilon)\leq EN_m^+(\varepsilon)+\sum_{k=1}^{m}\sum_{i=1}^{b(k)}P\big[|X_i|\geq L_k\big]+$$

$$+2\sum_{k=1}^{m}P\Big[|S^*_{a(k)}|\geq \frac{a(k)\varepsilon}{4}\Big]+\sum_{k=1}^{m}P\Big[\max_{a(k)\leq l\leq b(k)}|S^*_l|\geq \frac{a(k)\varepsilon}{4}\Big],$$

where the quantities used here have the same meaning as in above.

But now we have

$$P\Big[|S^*_{a(k)}|\geq \frac{a(k)\varepsilon}{4}\Big]\leq P\Big[|S'_{a(k)}|+|ES^{**}_{a(k)}|\geq \frac{a(k)\varepsilon}{4}\Big]\leq$$

$$\leq P\Big[|S'_{a(k)}|\geq \frac{a(k)\varepsilon}{4}-a(k)\sup_i E|X_i|I\big[|X_i|\geq L_k\big]\Big]\leq$$

$$\leq P\Big[|S'_{a(k)}|\geq \frac{a(k)\varepsilon}{4}(1-\delta)\Big]$$

for sufficiently large k as $\sup_i E|X_i|I[|X_i|\geq L_k]\to 0$ when $k\to\infty$

Analogously, we get

$$P\Big[\max_{a(k)\leq l\leq b(k)}|S^*_l|\geq \frac{a(k)\varepsilon}{4}\Big]\leq$$

$$\leq P\Big[\max_{a(k)\leq l\leq b(k)}|S^*_l-ES^*_l|\geq \frac{a(k)\varepsilon}{4}-$$

$$-(b(k)-a(k))\sup_i E|X_i|I\big[|X_i|\geq L_k\big]\Big]\leq$$

$$\leq P\Big[\max_{1\leq l\leq b(k)}|S'_l|\geq \frac{a(k)\varepsilon}{4}(1-\delta)\Big].$$

Applying now Lemma 1 and Lemma 2 as in the proof of Lemma 3 we obtain (12).

Theorem 2. Let $\{X_k, k \geq 1\}$ be a sequence of independent random variables with $EX_k = 0$, $EX_k^2 < C$, $k \geq 1$, where C is a positive constant and such that for any given $\varepsilon > 0$

$$\sum_{n=1}^{\infty} n \sup_k P[|X_k| \geq \varepsilon n] < \infty$$

If $\{\nu_k, k \geq 1\}$ and $\nu$ are the quantities of Lemma 4 such that for any given $\varepsilon > 0$ $EN_{\infty}^{+}(\varepsilon) < \infty$ then $EN_{\infty}(\varepsilon) < \infty$.

More precisely, for any given $\varepsilon > 0$ and a fixed $\delta \in (0, \frac{1}{3})$

$$\sum_{n=1}^{\infty} P[|S_{\nu_n}| \geq \varepsilon \nu_n] \leq K_2 \Big( \sum_{n=1}^{\infty} P[|\nu_n - n\nu| \geq n\varepsilon]$$

$$+ \sum_{n=1}^{\infty} n \sup_k P[|X_k| \geq \tfrac{1}{8}(c-\varepsilon)(1-3\delta)\varepsilon n] + \sum_{n=1}^{\infty} n^{-1-\frac{\delta}{1-2\delta}} \Big),$$

where $K_2$ is a positive constant.

Moreover, we have

$$\sum_{k=1}^{m} \left( \frac{\sum_{i=1}^{k} E|X_i|^2 I[|X_i| \leq \frac{1}{8}(c-\varepsilon)(1-3\delta)\varepsilon k]}{k^2} \right)^{1+\frac{\delta}{1-2\delta}}$$

$$\leq \sum_{k=1}^{m} \left(\frac{c}{k}\right)^{1+\frac{\delta}{1-2\delta}} < \infty$$

Corollary. Let $\{X_k, k \geq 1\}$ be a sequence of independent random variables with

$$EX_k = 0 \ , \ EX_k^2 \log^{1+\eta}(1+|X_k|) < C \ , \ k \geq 1,$$

where C and $\eta$ are positive constants.

If $\{\nu_k, k \geq 1\}$ and $\nu$ are quantities of Lemma 4 such that for any given $\varepsilon > 0$ $EN_{\infty}^{+}(\varepsilon) < \infty$ then $EN_{\infty}(\varepsilon) < \infty$.

More precisely, for any given $\varepsilon > 0$ and a fixed $\delta \in (0, \frac{1}{3})$

$$\sum_{n=1}^{\infty} P[|S_{\nu_n}| \geq \varepsilon \nu_n] \leq K_3 \max \Big\{ \sum_{n=1}^{\infty} P[|\nu_n - n\nu| \geq n\varepsilon],$$

$$\sum_{n=1}^{\infty} \frac{1}{n \ln^{1+\eta} n} \Big\},$$

where $K_3$ is a positive constant.

Proof. Let us observe that

$$\sum_{k=1}^{m} \sum_{i=1}^{3dk} P[|X_i| \geq \tfrac{1}{8}(c-\varepsilon)(1-3\delta)\varepsilon k] \leq$$

$$\leq 3d \sum_{k=1}^{n} k \sup_i P\left[|X_i| \geq \frac{1}{8}(c-\varepsilon)(1-3\delta)\varepsilon k\right] \leq$$

$$\leq M \sum_{k=1}^{n} \frac{1}{k \log^{1+\eta} k} \sup_i EX_i^2 \log^{1+\eta}(1+|X_i|) \leq$$

$$\leq MC \sum_{k=1}^{n} \frac{1}{k \log^{1+\eta} k} < \infty$$

Hence , using the assumptions of the Corollary and Lemma 4 we get $EN_\infty(\varepsilon) < \infty$.

To give an estimation of $EN_n(\varepsilon)$ in the case of another class of independent non-identically distributed random variables we need the following

Definition. A sequence $\{X_n, n \geq 1\}$ of random variables with $EX_n = 0$, $n \geq 1$ , is said to satisfy the condition (C) if there exists a random variable X and positive constants a , b and $x_0$ such that for all $n \geq 1$ and $x > x_0$

$$\text{(C)} \qquad n^{-1} \sum_{k=1}^{n} P\left[|X_k| \geq x\right] \leq a\, P\left[|X| > bx\right].$$

Lemma 5. Let $\{X_n, n \geq 1\}$ be a sequence of independent random variables with $EX_n = 0$ , $n \geq 1$ , satisfying the condition (C) and let $\{\nu_k, k \geq 1\}$ be a sequence of positive integer-valued random variables defined on the same probability space $(\Omega, \mathcal{A}, P)$ as $\{X_k, k \geq 1\}$.
Denote by $\nu$ a random variable taking values from the interval $(c,d)$, $0 < c < d < \infty$.
Then for any given $\varepsilon > 0$ and a fixed $\delta \in (0, \frac{1}{3})$

$$(14) \quad EN_n(\varepsilon) = O\Big(EN_n^+(\varepsilon) + \sum_{k=1}^{n} k\, P\left[|X| \geq \frac{1}{8}(c-\varepsilon)(1-3\delta) b \varepsilon k\right] + \sum_{k=1}^{n} \Big(\frac{1}{k} \sum_{j=1}^{k} j\, P\left[|X| > jb\varepsilon\right]\Big)^{1+\frac{\delta}{1-2\delta}}\Big)$$

Proof. Taking into account the condition (C) we see that there exists a positive constant M such that

$$\sum_{k=1}^{n} \sum_{i=1}^{3dk} P\left[|X_i| \geq \frac{1}{8}(c-\varepsilon)(1-3\delta)\varepsilon k\right] \leq$$

$$\leq M \sum_{k=1}^{n} k\, P\left[\,|X| \geq \tfrac{1}{8}(c-\varepsilon)(1-3\delta)b\varepsilon k\right].$$

Moreover , supposing that $c < 1$ / without loss of generality /, we note that

$$EX_i^2\, I\left[\,|X_i| \leq \tfrac{1}{8}(c-\varepsilon)(1-3\delta)\varepsilon k\right] \leq$$

$$\leq EX_i^2\, I\left[\,|X_i| < k\varepsilon\right] = -\int_0^{k\varepsilon} x^2\, dP\left[\,|X_i| > x\right] =$$

$$= -(k\varepsilon)^2 P\left[\,|X_i| \geq k\varepsilon\right] + 2\int_0^{k\varepsilon} x\, P\left[\,|X_i| > x\right] dx \leq$$

$$\leq 2\sum_{j=1}^{k} \int_{(j-1)\varepsilon}^{j\varepsilon} x\, P\left[\,|X_i| > x\right] dx.$$

Hence , we conclude that there exists a positive constant $C = C(a, \varepsilon)$ such that

$$\sum_{i=1}^{k} EX_i^2\, I\left[\,|X_i| < k\varepsilon\right] \leq 2\sum_{i=1}^{k}\sum_{j=1}^{k} \int_{(j-1)\varepsilon}^{j\varepsilon} x\, P\left[\,|X_i| > x\right] dx =$$

$$= 2\sum_{j=1}^{k}\sum_{i=1}^{k} \int_{(j-1)\varepsilon}^{j\varepsilon} x\, P\left[\,|X_i| > x\right] dx \leq Ck\sum_{j=1}^{k} \int_{(j-1)\varepsilon}^{j\varepsilon} x\, P\left[\,|X| > bx\right] dx \leq$$

$$\leq Ck \textstyle\sum_{j=1}^{k} j\, P\left[\,|X| > b(j-1)\varepsilon\right].$$

We note now that under the condition (C) (12) holds. Thus , using the estimates derived above , we obtain (14).

From Lemma 1 we get the following extension of results of [3] and [4] .

Theorem 3. If under the assumptions of Lemma 5 $EX^2 < \infty$ and for any given $\varepsilon > 0$ $EN^+_\infty(\varepsilon) < \infty$ then $EN_\infty(\varepsilon) < \infty$

More precisely , for any given $\varepsilon > 0$ and a fixed $\delta \in (0, \frac{1}{3})$

$$\sum_{n=1}^{\infty} P\left[\,|S_{\nu_n}| \geq \varepsilon \nu_n\right] \leq K_4\Bigg(\sum_{n=1}^{\infty} P\left[\,|\nu_n - n\nu| \geq n\varepsilon\right] +$$

$$+ \sum_{n=1}^{\infty} n\, P\left[\,|X| \geq \tfrac{1}{8}(c-\varepsilon)(1-3\delta)b\varepsilon n\right] +$$

$$+ \sum_{n=1}^{\infty} \left(\ln n / n\right)^{1+\frac{\delta}{1-2\delta}}\Bigg).$$

where $K_4$ is a positive constant.

## REFERENCES

[1] A.A. Borovkov , Remarks on inequalities for sums of independent variables. Teor. Verojatnost. i Primenen. 17(1972), 588 - 590.

[2] D.H. Fuk , S.V. Nogaev , Probability inequalities for sums of independent random variables. Teor. Verojatnost. i Primenen. 16 (1971), 643 - 660.

[3] P.L. Hsu , H. Robbins , Complete convergence and the low of large numbers. Proc. Nat. Acad. Sci. U.S.A. 33 (1947), 25 - 31.

[4] D. Szynal , On almost complete convergence for the sum of a random number of independent random variables. Bull. Acad. Polon. Sci., Ser. Math., Astronom., Phys. 20 (1972), 571 - 574.

# A LIMIT THEOREM FOR MARKOV RENEWAL PROCESSES

József Tomkó

University of Debrecen
Comp. and Automat. Inst. of the Hung. Acad. of Sci.

## 1. INTRODUCTION

The results on thinning of renewal processes ([6], [5], [8]) initiated an extension to the multivariate case that has led to an asymptotic analysis of Markov renewal processes [11]. Some queuing problems studied using semi-Markov processes give rise to limit theorems of [10], [7], [9], [12]. These theorems suggested that the asymptotic analysis of semi-Markov processes should be possible in more general situations. In a series of works ([1], [2], [3], [4]), Koroljuk and Turbin have considered fairly general situations for an asymptotic analysis of semi-Markov processes. Their tool, perturbation theory of linear operators, though being rather heavy, allows an asymptotic expansion for the limiting probability distribution. Slightly modifying the problem and using the ideas of [11], an asymptotic investigation is presented for more general situation than those of Koroljuk and Turbin. An example of such a limiting phenomenon is the limiting behaviour of busy cycles of an $E_k/G/1/N$ queue as the input intensity increases. This example is discussed in [13].

## 2. THE RESTRICTION OF A MARKOV RENEWAL PROCESS WITH RESPECT TO A SUBSET OF STATES

Let $\{t_k, r_k;\ k \geqslant 0\}$ be a Markov renewal process the second component of which has finite state space $E$. For simplicity suppose that $E$ is the set $\{1, 2, \ldots, N\}$, and $t_0 = 0$. Recall that for such a process $0 < t_1 < t_2 < \ldots < t_k < \ldots$, and for any integer $k \geqslant 0$, real $x > 0$ and $j \in E$ the conditional probability of the event

$$\{t_{k+1} - t_k \leqslant x,\ r_{k+1} = j\}$$

with respect to the whole past $\{t_s, r_s;\ 0 \leqslant s \leqslant k\}$ depends only on $r_k$. Let the transition functions, $G_{ij}(x)$, be defined as

$$G_{ij}(x) = \mathsf{P}\{t_{k+1} - t_k \leqslant x,\ r_{k+1} = j \mid r_k = i\} \qquad (i, j \in E,\ x > 0).$$

The process is characterized by the matrix

$$G(x) = (G_{ij}(x)) \qquad (i, j \in E). \tag{2.1}$$

Associated with the Markov renewal process is the semi-Markov process

$$\xi_t = r_{n(t)} \qquad (t \geqslant 0),$$

where $n(t) = \max\{k;\ t_k \leqslant t\}$. Note that the discrete valued process $\{\xi_t,\ t \geqslant 0\}$ is right continuous.

Instead of the matrix $G(x)$ we shall work with its Laplace – Stieltjes (L–S) transform $\Gamma(s)$ with elements $\Gamma_{ij}(s) = \int_0^\infty e^{-sx}\, dG_{ij}(x)$. The sequence $\{r_k,\ k \geqslant 0\}$ is a Markov chain with transition matrix

$$P = (p_{ij}) = (G_{ij}(\infty)) = (\Gamma_{ij}(0)).$$

Let $U$ be a subset of $E$. Suppose $r_0 \in U$ and consider the successive transition epochs $\{\tau_k,\ k \geq 0\}$ at which the process visits the set $U$. Define

$$\nu_k = \xi_{\tau_k} \qquad (k \geq 0),$$

$$\eta(t) = \max\{k;\ \tau_k \leq t\}$$

and

$$\xi_t^{(U)} = \nu_{\eta(t)} \qquad (t > 0).$$

The process $\{\tau_k, \nu_k;\ k \geq 0\}$ $(\{\xi_t^{(U)},\ t \geq 0\})$ is a Markov renewal (semi-Markov) process which will be called the restriction of $\{t_k, r_k;\ k \geq 0\}$ $(\{\xi_t,\ t \geq 0\})$ to the set $U$.

Denote by $\varphi(s)$, $\mathrm{Re}\, s \geq 0$, the matrix of L–S transform of transition probability distributions

$$F_{ij}(x) = \mathsf{P}\{\tau_{k+1} - \tau_k \leq x, \nu_{k+1} = j \mid \nu_k = i\} \qquad (i,j \in U). \tag{2.3}$$

To formulate the relation between $\varphi(s)$ and $\Gamma(s)$ consider the partition $E = U \cup V$ and write $\Gamma(s)$ in the following hypermatrix form

$$\Gamma(s) = \begin{bmatrix} \Gamma^{(U,U)}(s) & \Gamma^{(U,V)}(s) \\ \Gamma^{(V,U)}(s) & \Gamma^{(V,V)}(s) \end{bmatrix}$$

where for any two subsets $U, V$ of $E$

$$\Gamma^{(U,V)}(s) = (\Gamma_{ij}(s)) \qquad (i \in U,\ j \in V).$$

Obviously we have that

$$\varphi(s) = \Gamma^{(U,U)}(s) + \Gamma^{(U,V)}(s)[I - \Gamma^{(V,V)}(s)]^{-1}\Gamma^{(V,U)}(s). \tag{2.4}$$

It is quite easy to understand this formula. Starting from a state in $U$, the next transition may lead again to a state in $U$ which is the reason for the first term on the right hand side of (2.4). For the second term the reasoning is as follows. Starting from $U$ the next transition will lead to $V$ after which the process will remain in $V$ for several transitions and finally return to $U$.

It is hardly hopeful that (2.4) might be employed to obtain some deeper information on the distribution functions (2.3). But (2.4) is the key to studying the asymptotic behaviour of the restriction.

## 3. THE LIMIT THEOREM

Suppose that the subset $U$ is rarely visited by $\{\xi_t,\ t \geq 0\}$. Then we are concerned with a family of processes $\{\{\xi_t(\epsilon),\ t \geq 0\},\ \epsilon > 0\}$ governed by the family of matrices $\{\Gamma(\epsilon, s),\ \epsilon > 0\}$ or $\{G(\epsilon, x),\ \epsilon > 0\}$, such that the terms in $\Gamma^{(U,U)}(\epsilon, s)$ and $\Gamma^{(V,U)}(\epsilon, s)$ approach nil as $\epsilon \to 0$. It is intuitively clear that in such a case the restricted process $\{\xi_t^{(U)}(\epsilon),\ t \geq 0\}$ is rather thin in the sense that for small $\epsilon$ small numbers of transitions occur during any fixed finite time interval. Our goal is to describe the conditions under which the restriction suitably normalized converges to a limit process of simple structure. If one considers the initial Markov renewal process $\{t_k, r_k;\ k \geq 0\}$ as a multivariate point process, keeping in mind the results for rarefying point processes, the limit process appears to be Poissonian.

To list the conditions mentioned above, first consider the partition

$$E = E_n \cup E_{n-1} \cup \ldots \cup E_1 \cup E_0$$

and choose $U = E_n$, so that $V = \bigcup_{i=0}^{n-1} E_i$. To simplify notation, put

$$\Gamma^{(E_k, E_l)}(\epsilon, s) = \Gamma^{(k,l)}(\epsilon, s) \qquad (k, l = 0, 1, \ldots, n).$$

Our assumptions will be as follows:

(I) The matrix $\Gamma(\epsilon, 0)$ is irreducible for every $\epsilon > 0$.

(II) For $0 \leq k < l - 1 \leq n - 1$ $\Gamma^{(k,l)}(\epsilon, 0)$ are nil matrices.

(III) For $0 < k \leq n$ there exists

$$\lim_{\epsilon \to 0} \frac{1}{\epsilon} \Gamma^{(k-1,k)}(\epsilon, 0) = R^{(k)} = (r_{ij}^{(k)}) \qquad (i \in E_{k-1},\ j \in E_k).$$

(IV) For $0 \leq k \leq n - 1$ there exist

$$\lim_{\epsilon \to 0} \Gamma^{(k,k)}(\epsilon, 0) = G_k \quad \text{and} \quad \lim_{\epsilon \to 0} \Gamma^{(n,n)}(\epsilon, 0) = 0.$$

(V) For $\Gamma_i(\epsilon, s) = \sum_{j \in E} \Gamma_{ij}(\epsilon, s)$, if $i \in E_{n-1} \cup \ldots \cup E_1$, then $\dfrac{1 - \Gamma_i(\epsilon, s)}{s}$ is bounded for $\operatorname{Re} s > 0$, $\epsilon > 0$; if $i \in E_0$ then there exists $\Gamma_i'(\epsilon, 0)$ and $\lim_{\epsilon \to 0} \Gamma_i'(\epsilon, 0) = \mu_i$.

Roughly speaking $\Gamma(\epsilon, s)$ is a hypermatrix of the form

$$\text{(3.1)} \qquad \begin{bmatrix} \Gamma^{(n,n)}(\epsilon, s) & \Gamma^{(n,n-1)}(\epsilon, s) & \Gamma^{(n,n-2)}(\epsilon, s) & \cdots & \Gamma^{(n,0)}(\epsilon, s) \\ \Gamma^{(n-1,n)}(\epsilon, s) & \Gamma^{(n-1,n-1)}(\epsilon, s) & \Gamma^{(n-1,n-2)}(\epsilon, s) & \cdots & \Gamma^{(n-1,0)}(\epsilon, s) \\ 0 & \Gamma^{(n-2,n-1)}(\epsilon, s) & \Gamma^{(n-2,n-2)}(\epsilon, s) & \cdots & \Gamma^{(n-2,0)}(\epsilon, s) \\ \vdots & & & & \\ 0 & \cdots & 0 & \Gamma^{(0,1)}(\epsilon, s) & \Gamma^{(0,0)}(\epsilon, s) \end{bmatrix},$$

where the terms below the main diagonal are of order $\epsilon$. It is not difficult to quess that we should scale the restriction in accordance with

$$\tau = \epsilon^n t.$$

The scaled restriction is defined as

$$\text{(3.2)} \qquad \hat{\xi}_\tau^{(U)} = \xi_{\frac{\tau}{\epsilon^n}}^{(U)} \qquad (\tau > 0)$$

so that the L–S transform of the matrix governing it has the form

$$\varphi(\epsilon, \epsilon^n s) = \Gamma^{(n,n)}(\epsilon, \epsilon^n s) +$$

(3.3)

$$+ [\Gamma^{(n,n-1)}(\epsilon, \epsilon^n s) \dots \Gamma^{(n,0)}(\epsilon, \epsilon^n s)][I - A(\epsilon^n s)]^{-1} \begin{bmatrix} \Gamma^{(n-1,n)}(\epsilon, \epsilon^n s) \\ 0 \\ \vdots \\ 0 \end{bmatrix},$$

where $A(s)$ is obtained from (3.1) by omitting the blocks $\Gamma^{(n,k)}(\epsilon, s), \Gamma^{(k,n)}(\epsilon, s)$ $(0 \leqslant k \leqslant n)$. We formulate our

**Limit theorem.** *Suppose* $G_0$ *is ergodic and there exists a chain of states* $i_0, i_1, \dots, i_n$ *such that* $i_k \in E_k$ $(k = 1, 2, \dots, n)$ *and*

$$\prod_{n=1}^{n} r^{(k)}_{i_{k-1} i_k} > 0.$$

*Then under assumptions* (I), . . . , (V)

$$\lim_{\epsilon \to 0} \varphi(\epsilon, \epsilon^n s) = \frac{\lambda}{\lambda + s} P_0$$

*with some* $\lambda > 0$ *and* $P_0$ *is a matrix of identical rows.*

*In other words, the Markov renewal process* $\{\epsilon^n t_k, r_k;\ k \geqslant 0\}$ *converges to a multivariate Poisson process, or the semi-Markov process* $\{\hat{\xi}^{(U)}_\tau,\ \tau \geqslant 0\}$ *approaches a Markov chain with infinitesimal matrix* $a_{ii} = -\lambda$, $a_{ij} = \lambda p_j$, $1 \leqslant i, j \leqslant m$, *where* $p_j$ *is the j-th element of any row of* $P_0$, *and* $m$ *is the number of states in* $E_n$.

**Proof.** First we mention that as in Lemma 1 of [11] one verifies for $0 < k < n$

$$\lim_{\epsilon \to 0} \frac{1}{\epsilon} \Gamma^{(k-1,k)}(\epsilon, \epsilon^n s) = R^{(k)},$$

and for $i \in E_0$

$$\lim_{\epsilon \to 0} \frac{1 - \Gamma_i(\epsilon, \epsilon^n s)}{\epsilon^n s} = \mu_i. \tag{3.4}$$

Also note that $\lim_{\epsilon \to 0} \Gamma^{(n,n)}(\epsilon, \epsilon^n s) = 0$. Now examine $[I - A(\epsilon^n s)]^{-1}$. Write it in the form

$$\frac{\langle I - A(\epsilon^n s) \rangle}{|I - A(\epsilon^n s)|}$$

where, as in [11], $\langle B \rangle = (\bar{b}_{ij})$ with $(i,j)$-th cofactors of $B = (b_{ij})$. It can be shown that

$$\lim_{\epsilon \to 0} \epsilon^{-n} |I - A(\epsilon^n s)| = a + bs \tag{3.5}$$

with some $a > 0$, $b > 0$; and if $|q_{ij}|$ denotes the cofactors of $I - A(\epsilon^n s)$, then for $i \in E_{n-1}$ whenever $j \in V$, there exists

$$(3.6) \qquad \lim_{\epsilon \to 0} \frac{|q_{ij}|}{\epsilon^{n-1}} = q_i.$$

For the sake of simplification introduce $a_i = \sum_{j \in E_n} \Gamma_{ij}(\epsilon, \epsilon^n s) \quad (i \in E_{n-1})$,

$$h_{ij} = \Gamma_{ij}(\epsilon, \epsilon^n s) \qquad (i, j \in V).$$

In order to verify (3.5) modify the matrix $I - A(\epsilon^n s)$ by replacing the last column by the row sums. Thus the last column has elements $1 - \Gamma_i(\epsilon, \epsilon^n s) + a_i \chi_{\{i \in E_{n-1}\}}$ $(i \in V)$ ($\chi_C$ is the indicator of condition $C$). Therefore we have

$$|I - A(\epsilon^n s)| = |H_1| + |H_2|$$

where $H_1, H_2$ differ from $I - A(\epsilon^n s)$ only by their last column which consists of $1 - \Gamma_i(\epsilon, \epsilon^n s)$ $(i \in V)$ for $H_1$; and $a_i$ $(i \in E_{n-1})$, 0 otherwise, for $H_2$. Divide the last column in $H_1$ by $\epsilon^n$ and let $\epsilon \to 0$. We obtain

$$\lim_{\epsilon \to 0} \epsilon^{-n} |H_1| = sb = sK \prod_{k=1}^{n-1} d_k \sum_{i \in E_0} \pi_i \mu_i,$$

where $K$ is a constant depending only on $G_0$; $\pi_i$ $(i \in E_0)$ is the ergodic distribution of $G_0$; and $d_k = |I - G_k|$ $(0 < k < n)$ (for details see the arguments of Lemma 2 in [11]). To give a reason for the existence of

$$(3.7) \qquad a = \lim_{\epsilon \to 0} \epsilon^{-n} |H_2|$$

observe that $a_i$ $(i \in E_{n-1})$ are of order $\epsilon$ and that any term of the cofactors corresponding to the $a_i$ $(i \in E_{n-1})$ elements of the last column should contain at least one factor from every matrix, $\Gamma^{(k-1,k)}(\epsilon, \epsilon^n s)$ $(0 < k < n)$ which are of order $\epsilon$.

To show the existence of the limit (3.6), we first compare $q_{ij}$ with $q_{ij+1}$. Replacing the $j$-th column of $q_{ij+1}$ by its row sums, we have

$$|q_{ij+1}| = |R_1| + |q_{ij}| + |R_2|$$

where the $j$-th column, for $R_1$, is $1 - \Gamma_k(\epsilon, \epsilon^n s)$ $(k \in V,\ k \neq i)$; for $R_2$, is $a_k$ $(k \in E_{n-1},\ k \neq i)$, 0 otherwise; the other columns are the same as in $q_{ij}$. From assumption (V) and (3.4) it follows that

$$\epsilon^{-(n-1)} |R_1| \to 0 \quad \text{as} \quad \epsilon \to 0.$$

The fact that the same is true for $R_2$ can be shown by replacing its last column by its row sums, after which a decomposition similar to that considered above will lead to $R_1^*$ and $R_2^*$, where $R_1^*$ can be treated as $R_1$, $R_2^*$ as $H_2$ in case of $I - A(\epsilon, \epsilon^n s)$.

Consider now for $i \in E_{n-1}$ $q_{ii}$ and replace its last column by its row sums. Again decompose this modified version of $|q_{ii}|$ as $|D_1| + |D_2| + |D_3|$ with the last column $1 - \Gamma_k(\epsilon, \epsilon^n s)$ $(k \neq i,\ k \in V)$, for $D_1$; $h_{ki}$ $(k \neq i,\ k \in E_{n-1})$, 0 otherwise, for $D_2$; $a_k$ $(k \neq i,\ k \in E_{n-1})$, 0 otherwise in case of $D_3$. Similar arguments employed before, show that

$$\lim_{\epsilon \to 0} \epsilon^{-(n-1)} |D_1| = \lim_{\epsilon \to 0} \epsilon^{-(n-1)} |D_3| = 0$$

and that there exists

$$\lim_{\epsilon \to 0} \epsilon^{-(n-1)} |D_2| = q_i.$$

The positiveness of $q_i$ follows from $|q_{ii}| > 0$, which can be verified by induction.

Returning to the limit (3.7), we can easily see that

$$a = \lim_{\epsilon \to 0} \frac{1}{\epsilon^n} \sum_{i \in E_{n-1}} a_i |q_{ii}| = \sum_{i \in E_{n-1}} q_i \sum_{j \in E_n} r_{ij}^{(n)}.$$

In order to complete the proof, note that we have

$$\lim_{\epsilon \to 0} [\Gamma^{(n,n-1)}(\epsilon, \epsilon^n s) \ldots \Gamma^{(n,0)}(\epsilon, \epsilon^n s)] \frac{1}{\epsilon^{n-1}} \langle I - A(\epsilon, \epsilon^n s) \rangle \frac{1}{\epsilon} \begin{bmatrix} \Gamma^{(n,0)}(\epsilon, \epsilon^n s) \\ 0 \\ \vdots \\ 0 \end{bmatrix} =$$

$$= \begin{bmatrix} q_1 & q_2 & \cdots \\ q_1 & q_2 & \cdots \\ \vdots & \vdots & \cdots \\ q_1 & q_2 & \cdots \end{bmatrix} R^{(n)} = a \begin{bmatrix} p_1 & p_2 & \cdots & p_m \\ p_1 & p_2 & \cdots & p_m \\ \vdots & \vdots & \cdots & \cdot \\ p_1 & p_2 & \cdots & p_m \end{bmatrix}$$

with

$$p_i = \frac{\sum_{k \in E_{n-1}} q_k r_{ki}}{a} \qquad (1 \leq i \leq m). \tag{3.8}$$

Thus this remark and (3.5) proves the theorem with $\lambda = \frac{b}{a}$ and $p_i$ $(i \in E_n)$ given by (3.8).

REFERENCES


[1] V. Sz. Koroljuk – A.F. Turbin, On the asymptotic behaviour of SMP occupation times of the reducible set of states, *Theory of Prob. and Mat. Stat.*, 2 (1970), 133-144.

[2] V.Sz. Koroljuk – I.P. Penev – A.F. Turbin, Asymptotic expansion of absorbing time distribution for Markov chains, *Kibernetika,* 4 (1973), 133-135 (in Russian).

[3] V.Sz. Koroljuk – A.F. Turbin, *Semi Markov processes and their applications,* Naukova Dumka, Kiev, 1976 (in Russian).

[4] V.Sz. Koroljuk – A.F. Turbin, *Phase enlargement of complex systems,* Vyzsa Skóla, Kiev, 1978 (in Russian).

[5] J. Mogyoródi, On the rarefaction of renewal processes, I. II., *Studia Sci. Math. Hungarica,* 7 (1972), 285-291, 293-305.

[6] A. Rényi, A Poisson folyamat egy jellemzése, *MTA Mat. Kut. Int. Közleményei,* 1 (1956), 519-526 (in Hungarian).

[7] I.S. Rosenlund, Busy periods in time-dependent $M/G/1$ queues, *Adv. Appl. Prob.,* 8 (1976), 195-208.

[8] R.F. Serfőző, Composition, inverses and thinnings of random measures, *Z. Wahrscheinlichkeitstheorie verw. Geb.,* 37, No. 3 (1976/77), 253-265.

[9] D. Szász, A limit theorem for semi Markov processes, *J. Appl. Prob.,* 11 (1974), 521-528.

[10] J. Tomkó, Limit theorem for queueing system with infinitely increasing input intensity, *Studia Sci. Math. Hung.,* 2 (1967), 447-554 (in Russian).

[11] J. Tomkó, On the rarefaction of multivariate point processes, Colloquia Math. Soc. János Bolyai, 9, *Progress in statistics,* Budapest (Hungary), (1972), 834-860.

[12] J. Tomkó, Számológépek központi egységének kihasználtságáról, II, *Alkalmazott Matematikai Lapok,* 3 (1977), 83-96 (in Hungarian).

[13] J. Tomkó, Semi Markov analysis of an $E_k/G/1$ queue with finite waiting room, Colloquia Math. Soc. János Bolyai, 25, *Point processes and queuing problems,* Keszthely (Hungary) (to be published).

# REMARK TO THE DERIVATION OF THE CRAMÉR-FRÉCHET-RAO INEQUALITY IN THE REGULAR CASE

István Vincze

Mathematical Institut of the Hungarian
Academy of Sciences
Reáltanoda u.13-15.
BUDAPEST, HUNGARY H-1053

## Introduction

In the paper [10] a derivation of the $C$-$F$-$R$ inequality is given when the parameter space is completely arbitrary. The procedure is conveninent for handling both the regular case and the case of non-homogeneous set of densities; after general considerations the mentioned paper deals with the nonregular case only. The aim of the present paper is to point out that the method used in the mentioned work allows a derivation of the classical $C$-$F$-$R$ inequality under simpler conditions, imposed by the classical authors [2,4,5,8] or appeared in the textbooks [see eg.9.] . -Some remarks are made for the information quantity occuring in the variance bound too.

## 1. Derivation of a general inequality

Let $E_n$ and $B_n$ be the Euclidean $n$ -space and the $\sigma$-algebra of its Borel-subsets respectively. The random variable $X=(X_1,X_2,\ldots,X_n)$ -$X\in E_n$ - is distributed according to a law having density function $f(X;\vartheta)=f(X_1,X_2,\ldots,X_n,\vartheta)$ with respect to the $\sigma$ -finite measure $\mu$ ; here $\vartheta$ is an element of the parameter space the structure of which can be arbitrary (e.g. an abstract set).

Let us denote by

$$F=\{f(X;\vartheta);\ \vartheta\in\Theta\}$$

the set of the underlying densities for which we assume that $f(X;\vartheta)$ is completely determined by $\vartheta$ . Suppose that the real valued injective function $g(\vartheta)$ has the unbiased estimator $t(X)$:

$$E_{\vartheta}(t(X)) = g(\vartheta), \quad \vartheta \in \Theta \quad .$$

For the support of the densities the notation

$$A_{\vartheta} = \{X : f(X;\vartheta) > 0\}$$

will be used.

For our purposes the following set of densities will be introduced let us fix two elements $\vartheta$, $\vartheta'$ of $\Theta$ and let $\alpha$ be a real parameter

$$F_{\alpha} = \{f_{\alpha}(X;\vartheta,\vartheta') = (1-\alpha)f(X;\vartheta) + \alpha f(X;\vartheta'), \quad 0<\alpha<1\}.$$

It is clear that if we consider $\alpha \in (0,1)$ as unknown parameter, then we have a homogeneous set of densities with the support

$$A_{\vartheta,\vartheta'} = A_{\vartheta} \cup A_{\vartheta'} \quad ,$$

not depending on $\alpha$.

It can be obtained by direct calculation that the statistic

$$\hat{\alpha} = \frac{t(X) - g(\vartheta)}{g(\vartheta') - g(\vartheta)}$$

is an unbiased estimator for $\alpha$ :

$$(1.1) \qquad E_{\alpha}(\hat{\alpha}) = \alpha \,, \qquad 0<\alpha<1 \,.$$

<u>For the variance of</u> $\hat{\alpha}$ <u>the following relation holds:</u>

$$(1.2) \qquad D_{\alpha}^{2}(\alpha) \geq \frac{1}{\int\limits_{A_{\vartheta,\vartheta'}} \frac{[f(X;\vartheta')-f(X,\vartheta)]^2}{f_{\alpha}(X;\vartheta,\vartheta')}\, d\mu(X)} \overset{\text{def}}{=} \frac{1}{I_{\alpha}(\vartheta,\vartheta')} \quad .$$

Relation (1.2) agrees with the classical $C$-$F$-$R$ inequality in the special case, when the density function depends linearly on the parameter $\alpha$ . For the sake of completeness we turn to the proof:

Let $0<\alpha$ , $\alpha+\Delta<1$ , then according to (1.1)

$$\alpha+\Delta = \int_{A_{\vartheta,\vartheta'}} \hat{\alpha} f_{\alpha+\Delta}(x;\ \vartheta,\vartheta')\ d\mu(x)$$

and

$$\alpha = \int_{A_{\vartheta,\vartheta'}} \hat{\alpha} f_{\alpha}(x;\ \vartheta,\vartheta')\ d\mu(x)\ .$$

Further

$$\alpha \int_{A_{\vartheta,\vartheta'}} [f_{\alpha+\Delta}(x;\vartheta,\vartheta') - f_{\alpha}(x;\vartheta,\vartheta')]\ d\mu(x) = 0$$

and

$$f_{\alpha+\Delta}(x;\vartheta,\vartheta') - f_{\alpha}(x;\vartheta,\vartheta') = \Delta(f(x;\vartheta') - f(x;\vartheta))\ .$$

Hence

$$1 = \int_{A_{\vartheta,\vartheta'}} (\hat{\alpha}-\alpha)\ (f(x;\vartheta') - f(x;\vartheta))\ d\mu(x) =$$

$$= \int_{A_{\vartheta,\vartheta'}} (\hat{\alpha}-\alpha)\ [f_{\alpha}(x;\vartheta,\vartheta')]^{\frac{1}{2}}\ \frac{f(x,\vartheta')-f(x,\vartheta)}{f_{\alpha}(x,\vartheta,\vartheta')}[f_{\alpha}(x,\vartheta,\vartheta')]^{\frac{1}{2}}\ d\mu(x).$$

The Bunjakowski-Cauchy-Schwarz inequality completes the proof. -

(1.2) can be written in the form

$$(1-\alpha)\ D^2_{\vartheta}(t(x)) + \alpha\ D^2_{\vartheta'}(t(x)) \geq (g(\vartheta') - g(\vartheta))^2\{\ \frac{1}{I_{\alpha}(\vartheta,\vartheta')} - \alpha(1-\alpha)\}$$

which is the known bound for the mixed variance. (See e.g. [6] ).

## 2. The homogeneous case

We turn now to the homogeneous case, i.e. let the support $A_{\vartheta}=A$ be independent of $\vartheta$; all relations being valid for $0\leq\alpha\leq 1$ we have

$$(2.1) \qquad D_\vartheta^2(t(x)) \geq \sup_{\vartheta'} \frac{(g(\vartheta') - g(\vartheta))^2}{\int_A \frac{[f(x;\vartheta')-f(x;\vartheta)]^2}{f(x;\vartheta)} d\mu(x)}$$

which corresponds to the $C$-$F$-$R$ inequality for an arbitrary parameter set.

Let now $\vartheta$ be real, $\Theta$ an open interval; assuming that $g(\vartheta)$ and $f(x;\vartheta)$ are differentiable with respect to $\vartheta$, we can turn to the limiting case when $\vartheta' \to \vartheta$ -assuming that the quantities occuring exist -

$$(2.2) \qquad D_\vartheta^2(t(x)) \geq \frac{[g'(\vartheta)]^2}{\int_A \frac{1}{f(x;\vartheta)} \left(\frac{\partial f(x;\vartheta)}{\partial \vartheta}\right)^2 d\mu(x)}$$

having the classical form of the $C$-$F$-$R$ inequality.

As shown by Chapman and Robbins [3] the form (2.1) even in the case of a real valued $\vartheta$ -assuming the differentibility of $g(\vartheta)$ and $f(\cdot\,;\vartheta)$ -may give a better bound for the variance.
But we emphasize that in our procedure no assumption concerning exchangebility of differentiation and integration was made.

## 3. Some remarks to the information quantity

In relations (1.2) and (2.1) the information quantity $I_\alpha(\vartheta,\vartheta')$ occurs-depending on two distinct elements of the parameter set $\Theta$. In the homogeneous case when $\vartheta$ is real and the differentiability of $f(\cdot\,,\vartheta)$ is assumed, then we have the relation

$$\frac{1}{(\vartheta'-\vartheta)^2} I_\alpha(\vartheta,\vartheta') \to \int_A \frac{1}{f(x;\vartheta)} \left(\frac{\partial f(x;\vartheta)}{\partial \vartheta}\right)^2 d\mu(x) ,$$

i.e. we come to Fisher-information. In his paper [11] the author showed that for $\alpha=1/2$, $I_{1/2}(\vartheta,\vartheta')$ can be considered as a natural measure for the information contained in a sample concerning the question whether $\vartheta$ or $\vartheta'$ turns out to be the true value of the parameter.

In our joint work with M.L.Puri [7], the following relations are proved for $I_\alpha(\vartheta,\vartheta')$ :

$$I_\alpha(\vartheta,\vartheta') \leq \frac{1}{\alpha(1-\alpha)}$$

and equality holds in the only case when the supports of the two densities are disjoint $(\mu)$.

Let us denote by $\varphi_\alpha(\vartheta,\vartheta')$ the Hellinger transform $(o \leq \alpha \leq 1)$

$$\varphi_\alpha(\vartheta,\vartheta') = \int_{A_\vartheta \cap A_{\vartheta'}} [f(x;\vartheta)]^\alpha [f(x;\vartheta')]^{1-\alpha} \, d\mu(x) ,$$

which was introduced for $\alpha = 1/2$ by A. Bhattacharyya [1]. Let further $h(\vartheta,\vartheta')$ be the Hellinger-distance:

$$h^2(\vartheta,\vartheta') = \int_{A_{\vartheta,\vartheta'}} [\sqrt{f(x;\vartheta')} - \sqrt{f(x;\vartheta)}]^2 \, d\mu(x) =$$

$$= 2[1-\varphi_{\frac{1}{2}}(\vartheta,\vartheta')]$$

Then the following relations are valid:

$$(3.1) \qquad \alpha(1-\alpha)\, I_\alpha(\vartheta,\vartheta') \leq 1+(1-\alpha)\, P_\vartheta(A_\vartheta \cap A_{\vartheta'}) + \alpha P_{\vartheta'}(A_\vartheta \cap A_{\vartheta'}) - 2\varphi_{\frac{1}{2}}(\vartheta,\vartheta') ,$$

$$(3.2) \qquad 1 \geq \frac{1}{4} I_{\frac{1}{2}}(\vartheta,\vartheta') \geq 1-\varphi_{1/2}(\vartheta,\vartheta') .$$

Considering relation (3.1) for $\alpha = o$ and $\alpha = 1$ we obtain

$$2\varphi_{\frac{1}{2}}(\vartheta,\vartheta') \leq 1+\min[P_\vartheta(A_\vartheta \cap A_{\vartheta'}),\ P_{\vartheta'}(A_\vartheta \cap A_{\vartheta'})] .$$

For some consequences of these relations and for further investigation of the $C$-$F$-$R$ inequality in the nonregular case the author intends to return in a joint paper with M.L.Puri.

## References


1. Bhattacharyya, A.: On a measure of divergence between two statistical populations defined by their probability distributions. Calcutta Mathematical Bulletin-35 (1943) 99-109.
2. Barankin, E.W. : Locally best unbiased estimator. Ann.Math.

Statist. 20(1949) 477-501.

3. Chapman, D.C.-H.Robbins: Minimum variance estimation without regularity conditions. Ann.Math.Statist. 22 (1951) 581-586.

4. Cramèr, H.: Mathematical Methods of Statistics, Princeton University Press. Princeton 1946.

5. Fréchet, M.: Sur l'extension de certaines evaluations statistiques au cas de petit echantillons. Rev. Inst. Internat. Statist. 11(1943) 182-205.

6. Polfeldt,T.: Asymtotic Results in Non-Regular Estimation. Skandinavisk Aktuarietidskrift.1970. 1-2. Supplement

7. Puri M.L.-I.Vincze: On the Cramér- Fréchet-Rao inequality for translation parameter in the case of finite support. (to be published)

8. Rao,C.R.: Information and accuracy attainable in the estimation of statistical parameters. Bull. Calcutta Math. Soc. 37(1945)81-91.

9. Schmetterer, L.: Einführung in die Mathematische Statistik. Springer. New-York-Wien 1966.

10. Vincze,I.: On the Cramér-Fréchet-Rao inequality in the non-regular case. Contributions to Statistics. Hájek Memorial Volume. Academia, Prague. 1979.p. 253-262.

11. Vincze,I.: On concept and measure of information contained in an observation (to be published).

# NONPARAMETRIC DENSITY ESTIMATORS IN ABSTRACT AND HOMOGENEOUS SPACES

Wolfgang Wertz

University of Technology, Wien

Summary: The problem of estimating probability densities by generalized kernel estimators, based on independent observations in an abstract space is investigated. Some asymptotic results (consistency, asymptotic unbiasedness) are given and applications to special spaces X (first of all homogeneous spaces) are considered.

## Introduction

Since the appearance of Rosenblatt's paper [7] in 1956, a great number of papers on the problem of density estimation and related topics has appeared (see e.g. *Schneider* and *Wertz* [9]). Most of the authors consider problems with observations on the real line or in an Euclidean space. But for many practical problems homogeneous spaces are appropriate sample spaces, in geology, for example, the sphere $S_2 = SO_3/SO_2$; an excellent survey on statistical problems in geology is given by *Watson* [12]. A more recent reference is *Mardia* and *Gadsen* [4]. At present, several statistical investigations with observations in homogeneous spaces are available, e.g. *Beran* [1], *Rukhin* [8] and *Wertz* [13].

## Density estimation in abstract spaces

Let X be a locally compact topological space, $\mathfrak{X}$ the $\sigma$-algebra of its Borel-sets, $\mu$ a Borel-measure on $(X,\mathfrak{X})$, $\mathcal{F}$ a class of bounded probability density functions with respect to $\mu$, $(\xi_n)$ a sequence of independent, identically distributed X-valued random variables, each one distributed with probability density

$f \in \mathcal{F}$. We consider (generalized) kernel estimators of f of the form:

$$\hat{f}_n(x_1,\dots,x_n;x) := \sum_{i=1}^{n} K_n(x,x_i),$$

where $(n,x,y) \mapsto K_n(x,y)$ is a sequence of kernels satisfying:

(1.1) $(x,y) \mapsto K_n(x,y)$ is $\mathfrak{X} \otimes \mathfrak{X}$-measurable for every n;

(1.2) $\int_X K_n(x,y)\, d\mu(y) = 1$ for every $x \in X$;

(1.3) $K_n \geqslant 0$;

(1.4) $\int_{\complement U} K_n(x,y)\, d\mu(y) \to 0$ as $n \to \infty$ for every $x \in X$ and every neighbourhood U of x;

and, occasionally,

(1.5) $\| K_n(x,\cdot)\|_\infty =: s_n(x) < \infty$ for every $x \in X$, $\|\cdot\|_\infty$ denoting the essential supremum-norm with respect to $\mu$.

In the following we shall make use of Hoeffding's inequality:

Lemma 1 ([3]): *Let* $\eta_1,\dots,\eta_n$ *be independent, identically distributed, bounded, W-integrable real random variables, say* $|\eta_i| \leqslant c$ *and* $E\eta_i = a$. *Then, for every* $t \in (0,c-a)$

$$W[|\frac{1}{n} \sum_{i=1}^{n} \eta_i - a| \geqslant t] \leqslant 2 \exp(-nt^2/2c^2)$$

*holds.*

Proposition 1: *Let* $(\hat{f}_n)$ *be a sequence of kernel estimators with the kernel fulfilling (1.1) - (1.4), and* $x \in X$. *Then for every* $f \in \mathcal{F}$, *which is continuous at* x,

$$\lim_{n\to\infty} E_f\, \hat{f}_n(\xi_1,\dots,\xi_n;x) = f(x)$$

*holds, i.e.:* $(\hat{f}_n)$ *is asymptotically (expectation-)unbiased.*

The *proof* is standard: Given an $\varepsilon > 0$, there is a neighbourhood $U_\varepsilon$ of $x$ with $|f(y) - f(x)| < \varepsilon$ for every $y \in U_\varepsilon$. Then

$$|E_f\ \hat{f}_n(\xi_1,\ldots,\xi_n;x) - f(x)| = |\int K_n(x,y).f(y)d\mu(y) -$$

$$- f(x).\int K_n(x,y)d\mu(y)| = |\int K_n(x,y).[f(y) - f(x)]\ d\mu(y)| \leqslant$$

$$\leqslant \int_{U_\varepsilon} \varepsilon.K_n(x,y)\ d\mu(y) + 2.\|f\|_\infty.\int_{\complement U_\varepsilon} K_n(x,y)\ d\mu(y) \leqslant$$

$$\leqslant \varepsilon + 2.\|f\|_\infty.\int_{\complement U_\varepsilon} K_n(x,y)\ d\mu(y).$$

The second term tends to zero as n tends to infinity, hence the assertion is proved.

q.e.d.

<u>Theorem 1:</u> *Let the assumptions of proposition* 1 *and* (1.5) *be satisfied. Further let* $\lim_{n\to\infty} n.s_n^{-2}(x) = \infty$, $f \in \mathcal{F}$ *be continuous at* $x$, $\mu$ *outer regular and* $\mu(\{x\}) = 0$. *Then* $(\hat{f}_n)$ *is locally consistent at* $x$, *more precisely: for every* $\varepsilon > 0$ *and* $n \geqslant n_0(\varepsilon,f,x)$,

$$W_f[|\hat{f}_n(\xi_1,\ldots,\xi_n;x) - f(x)| \geqslant \varepsilon] \leqslant 2.\exp[-n\varepsilon^2/8s_n^2(x)]$$

*holds. If moreover*

$$\sum_{n=1}^{\infty} \exp[-\delta n/s_n^2(x)] < \infty \quad \textit{for every} \quad \delta > 0 \tag{3}$$

*holds, then* $(\hat{f}_n)$ *is locally strongly consistent at*

$$W_f[\lim_{n\to\infty} \hat{f}_n(\xi_1,\ldots,\xi_n;x) = f(x)] = 1.$$

*Proof:* First of all, by (1.2), $\int K_n(x,y).f(y)\ d\mu(y) \leqslant \leqslant \|f\|_\infty < \infty$. We assert that $\lim_{n\to\infty} s_n(x) = \infty$. Otherwise, there is a bounded subsequence of $s_n(x)$, hence there is a $c < \infty$ and a sequence $(n_k)$ of natural numbers, $n_k \to \infty$, with $\|K_{n_k}(x,\cdot)\|_\infty \leqslant c$ for every k and

$$\int_U K_{n_k}(x,y)\ d\mu(y) \leqslant c.\mu(U)$$

for every neighbourhood U of X. By the regularity of $\mu$, tne condition $\mu(\{x\}) = 0$ and the local compactness of X, $c.\mu(U)$ can be made arbitrary small by suitable choice of U. $k \to \infty$ yields a contradiction to (1.4).

It follows that $\lim_{n\to\infty}[s_n(x) - \int K_n(x,y).f(y)\,d\mu(y)] = \infty$, hence Hoeffding's inequality applies for sufficiently large n:

$$W_f[|\frac{1}{n}\sum_{i=1}^{n} K_n(x,\xi_i) - E_f K_n(x,\xi)| \geq \varepsilon/2] \leq 2.\exp[-n\varepsilon^2/8s_n^2(x)].$$

Together with Proposition 1 this gives local consistency.
Now,

$$W_f[\hat{f}_n(\xi_1,\ldots,\xi_n;x) \not\to f(x)] \leq \sum_{k=1}^{\infty} W_f(\bigcap_{n=1}^{\infty}\bigcup_{l=n}^{\infty} [|\hat{f}_n(\xi_1,\ldots,\xi_n;x) -$$

$$- f(x)| \geq 1/k]),$$

but the event in parentheses can be written

$$\text{(4)}\quad \limsup_{n\to\infty}[|\hat{f}_n(\xi_1,\ldots,\xi_n;x) - f(x)| \geq 1/k].$$

The inequality

$$\sum_{n=n_0}^{\infty} W_f[|\hat{f}_n(\xi_1,\ldots,\xi_n;x) - f(x)| \geq 1/k] \leq$$

$$\leq 2.\sum_{n=n_0}^{\infty} \exp[-n/8k^2 s_n^2(x)]$$

is valid for sufficiently large $n_0$, and the right hand side is finite by hypothesis; by the Borel-Cantelli lemma, for every k, the event (4) has $W_f$-measure 0 and the strong consistency follows.

q.e.d.

Remark 1: A condition, similiar to (3) appears in papers of *Nadaraya* ([5]) and of *Schuster* ([10], [11]), dealing with kernel estimators on the real line of the form

$$\hat{f}_n(x_1,\ldots,x_n;x) = (1/nb_n).\sum_{i=1}^{n} K(\frac{x-x_i}{b_n}).$$

On the one hand, Nadaraya and Schuster obtain and characterize convergence of $(\hat{f}_n)$ towards f, uniform in x, using inequalities for the empirical distribution function, which are only available on the real line; it cannot be expected to get also uniform strong consistency in our general situation without stringent further assumptions. On the other hand, (1.1) - (1.5) are weaker than Nadaraya's and Schuster's condition that K is a probability density of bounded variation. At least uniform asymptotic unbiasedness can be proved, if X is a uniform space:

Corollary 1: *Let X be a uniform space and* $\mathcal{W}$ *a base for the uniform structure. Let* (1.1) - (1.3) *and additionally*

$$(5)\qquad \lim_{n\to\infty} \sup_{x\in X} \int_{\complement V(x)} K_n(x,y)\, d\mu(y) = 0 \quad \textit{for every uniformity } V \in \mathcal{W}$$

*be valid.* ($V(x)$ *denotes the* x*-section of* V).
*Then* $\lim_{n\to\infty} \sup_{x\in X} |E_f \hat{f}_n(\xi_1,\dots,\xi_n;x) - f(x)| = 0$ *for every uniformly continuous, bounded* f.

Remark 2: In the case X = G = locally compact group, μ (left invariant) Haar measure on X, kernels can be defined by $K_n(x,y) := T_n(x^{-1}y)$, where $T_n$ is a probability density with respect to μ, satisfying

$$\lim_{n\to\infty} \int_{\complement U} T_n\, d\mu = 0 \text{ for every } U \in \mathfrak{U}_e,$$

$\mathfrak{U}_e$ denoting a neighbourhood base of the unit element e of the group. Then (5) is fulfilled, because the sets of the form $V = \{(x,y) \in X \times X : x^{-1}y \in U\}$ ($U \in \mathfrak{U}_e$) define a uniform structure on X and

$$\int_{\complement V(x)} K_n(x,y)\, d\mu(y) = \int_{\complement xU} T_n(x^{-1}y)\, d\mu(y) = \int_{\complement U} T_n d\mu \to 0$$

for every $U \in \mathfrak{U}_e$.

In certain situations, the uniform continuity of f implies its boundedness, and the latter condition can be dropped in Corollary 1:

Lemma 2: *Let X be a uniform space with Borel sets* $\mathfrak{X}$, μ *a Borel measure on* $\mathfrak{X}$ *and* f *a uniformly continuous density with respect to* μ. *For every* V *belonging to a base* $\mathcal{W}$ *let* $\mu(V(x))$ *be independent of* x. *Then* f *is bounded.*

*The last condition is valid in every locally compact group X with left resp. right Haar measure and $\mathcal{V}$ a base for the left resp. right uniform structure.*

*Proof:* If f were not bounded, a sequence $x_0, x_1, x_2, \ldots$ in X would exist with $f(x_0) > 0, f(x_1) > f(x_0) + 1, \ldots, f(x_n) > f(x_{n-1}) + n$. For a fixed $\varepsilon \in (0,1/2)$ there is a uniformity $V = V_\varepsilon \in \mathcal{V}$ with $|f(x) - f(y)| < \varepsilon$ for every $(x,y) \in V$. The inequality

$$1 = \int f \, d\mu \geqslant \sum_{n=1}^{\infty} \int_{V(x_n)} [f(x_n) - \varepsilon] \, d\mu(x) \geqslant$$

$$\geqslant \sum_{n=1}^{\infty} \int_{V(x_n)} (n-\varepsilon) \, d\mu(x) = \mu(V(x_n)) . \sum_{n=1}^{\infty} (n-\varepsilon) = \infty$$

yields a contradiction.

In a topological group, the left uniform structure is defined as in Remark 2. For every $U \in \mathcal{U}_e$, $V(x) = xU$, if V is defined by U. By left invariance of Haar measure, $\mu(V(x)) = \mu(xU) = \mu(U)$ for every x. The case of right uniform structure is analogous.

q.e.d.

## Estimation in homogeneous spaces

Let X be a *homogeneous space* of the form $X = G/H$ with a locally compact $T_0$-group and a compact subgroup H. D denotes the modular function of G, the modular function of H is identically 1. Since the function $h \mapsto 1/D(h)$ $(h \in H)$ clearly extends to the whole of G as a continuous homomorphism $\chi : G \to (0,\infty)^x$, $(0,\infty)^x$ being the multiplicative group of the positive real numbers, there is a relatively invariant measure $\mu$ on $(X, \mathfrak{X})$ (see *Gaal* [2], Theorem V.3.12), more precisely:

$$\mu(gA) = \chi(g).\mu(A) \quad \text{for every } g \in G \text{ and } A \in \mathfrak{X} .$$

In general, the extension $\chi$ of $1/D$ from H is not unique; if G is not unimodular and $H = \{e\}$, then $\chi_1 \equiv 1$ and $\chi_2 \equiv 1/D$ are different extensions.

Let $\nu$ be a Haar measure on G and $\kappa$ a properly normed Haar measure of H. Then for every nonnegative $\nu$-integrable function $f: G \to \mathbb{R}$ the following equation holds:

$$(6) \qquad \int_G f\, d\nu = \int_{G/H} \int_H \frac{f(gh)}{\chi(gh)}\, d\kappa(h)\, d\mu(x).$$

(See *Reiter* [6], 8.2.3). The inner integral depends only on the coset $x = gH$ and not on $g$.
Let $(T_n)$ be a sequence of functions, $T_n : G \to \mathbb{R}$ with the following properties:

(7.1) $T_n(hg) = T_n(g)$ *for every* $h \in H$ *and* $g \in G$ *(in other words,* $T_n$ *is constant on every right coset);*

(7.2) $T_n$ *is a probability density with respect to* $\nu$*;*

(7.3) $\lim_{n\to\infty} \int_{\complement HUH} T_n(g)\, d\nu(g) = 0$ *for every* $U \in \mathfrak{U}_e$.

(HUH denots the set $\{huh' : u \in U;\ h,h' \in H\}$; (7.3) is weaker than the corresponding condition in Remark 2)
We take the following kernels:

$$(8) \qquad K_n(x,y) := \int_H \frac{T_n(g^{-1}g_1h_1)}{\chi(g_1h_1)}\, d\kappa(h_1) \text{ with } g \in x \text{ and } g_1 \in y.$$

It is easily seen that the integral is independent of the special choice of $g \in x$.

Remark 3: *Taking* $H = \{e\}$, *the extension* $\chi \equiv 1$ *of* $h \mapsto 1/D(h)$ *yields Haar measure* $\mu$. *Hence the estimators discussed in Remark 2 turn out to be a special case of the present definition.*

Proposition 2: *In case only the conditions* (7.1) *and* (7.2) *are fulfilled, the kernel estimators* $\hat{f}_n$ *with kernel* (8) *satisfy:*
$\hat{f}_n$ *is* $\chi$*-invariant (in the sense of Wertz* [13]*), that is*
$\hat{f}_n(x_1,\dots,x_n;x) = \chi(\gamma).\hat{f}_n(\gamma x_1,\dots,\gamma x_n;\gamma x)$ *for every* $x_i, x \in X$ *and* $\gamma \in G$. *If* $\chi \equiv 1/D$ *is used as an extension of* $\mathrm{Rest}_H(1/D)$ *to* G, *then* $x \to \hat{f}_n(x_1,\dots,x_n;x)$ *is a probability density with respect to* $\mu$ *for every* $(x_1,\dots,x_n) \in X^n$.

*Proof:*

$$\hat{f}_n(\gamma x_1,\ldots,\gamma x_n;\gamma x) =$$

$$= \frac{1}{n} \cdot \sum_{i=1}^{n} \int_H \frac{T_n((\gamma g)^{-1}\gamma g_i h_i)}{\chi(\gamma g_i h_i)} \, d\kappa(h_i) =$$

$$= \frac{1}{n} \cdot \sum_{i=1}^{n} \frac{1}{\chi(\gamma)} \int_H \frac{T_n(g^{-1}g_i h_i)}{\chi(g_i h_i)} \, d\kappa(h_i) =$$

$$= \frac{1}{\chi(\gamma)} \cdot \hat{f}_n(x_1,\ldots,x_n;x)$$

Now suppose $\chi \equiv 1/D$. Then by repeated application of Fubini's theorem, the properties of Haar integrals and (6) we get

$$\int_X \hat{f}_n(x_1,\ldots,x_n;x)\, d\mu(x) = \int_{G/H} \int_H \frac{T_n(g^{-1}g_1 h_1)}{\chi(g_1 h_1)} \, d\kappa(h_1)\, d\mu(x) =$$

$$= \frac{1}{\kappa(H)} \int_{G/H} \int_H \int_H \frac{T_n(h^{-1}g^{-1}g_1 h_1)\cdot\chi(gh)}{\chi(g_1 h_1)\cdot\chi(gh)} \, d\kappa(h)\, d\kappa(h_1)\, d\mu(x) =$$

$$= \frac{1}{\kappa(H)} \int_H \frac{1}{\chi(g_1 h_1)} \int_{G/H} \int_H \frac{T_n((gh)^{-1}g_1 h_1)\cdot\chi(gh)}{\chi(gh)} d\kappa(h) d\mu(x) d\kappa(h_1) =$$

$$= \frac{1}{\kappa(H)} \int_H \frac{1}{\chi(g_1 h_1)} \int_G T_n(g^{-1}g_1 h_1)\cdot\chi(g)\, d\nu(g)\, d\kappa(h_1) =$$

$$= \frac{1}{\kappa(H)} \int_H \int_G \frac{T_n(g^{-1}g_1 h_1)}{\chi(g^{-1}g_1 h_1)} \, d\nu(g)\, d\kappa(h_1) =$$

$$= \frac{1}{\kappa(H)} \int_H \int_G \frac{T_n(g^{-1})}{\chi(g^{-1})} \, d\nu(g)\, d\kappa(h_1) =$$

$$= \int_G \frac{T_n(g)}{\chi(g)} \cdot D(g^{-1})\, d\nu(g) = \int_G T_n(g)\, d\nu(g) = 1.$$

q.e.d.

Corollary 2: *Let (7.1) - (7.3) be valid and* ($\hat{f}_n$) *as in Proposition 2. Then* ($\hat{f}_n$) *is asymptotically unbiased at every point of continuity of* f.

*Proof:* By Proposition 1, the validity of (1.1) to (1.4) is to be established. The measurability of $K_n$ follows from general measure theory on groups (see *Reiter* [6], 8.2.3). (1.2) follows from

$$\int_X K_n(x,y)\, d\mu(y) = \int_{G/H}\int_H \frac{T_n(g^{-1}g_1h_1)}{\chi(g_1h_1)}\, d\kappa(h_1)\, d\mu(y) =$$

$$= \int_G T_n(g^{-1}g_1)\, d\nu(g_1) = \int_G T_n(g_1)\, d\nu(g_1) = 1,$$

(1.3) is trivial. To prove (1.4) let $\mathfrak{U}_e$ be a neighbourhood base of $e\varepsilon G$, consisting of symmetric neighbourhoods only, and $\mathfrak{W}$ the base of the uniformity of X consisting of sets V of the form

$$V := \{(x,y)\ \varepsilon\ X \times X:\ x^{-1}y \cap U \neq \emptyset\}$$

with $U\ \varepsilon\ \mathfrak{U}_e$, where $x^{-1}y := \{g^{-1}g_1:\ g\ \varepsilon\ x,\ g_1\ \varepsilon\ y\}$. Then $V(x) = gHUH = \{(ghu)H:\ h\ \varepsilon\ H,\ u\ \varepsilon\ U\}$ and

$$\int_{\complement V(x)} K_n(x,y)\, d\mu(y) = \int_{\complement V(x)}\int_H \frac{T_n(g^{-1}g_1h_1)}{\chi(g_1h_1)}\, d\kappa(h_1)\, d\mu(y) =$$

$$= \int_{G/H}\int_H I_{\complement V(x)}(y).\frac{T_n(g^{-1}g_1h_1)}{\chi(g_1h_1)}\, d\kappa(h_1)\, d\mu(y) =$$

$$= \int_{G/H}\int_H I_{\complement gHUH}(g_1h_1).\frac{T_n(g^{-1}g_1h_1)}{\chi(g_1h_1)}\, d\kappa(h_1)\, d\mu(y) =$$

$$= \int_G I_{\complement gHUH}(g_1).T_n(g^{-1}g_1)\, d\nu(g_1) =$$

$$= \int_G I_{\complement HUH}(g^{-1}g_1).T_n(g^{-1}g_1)\, d\nu(g_1) = \int_{\complement HUH} T_n(g_1)\, d\nu(g_1)$$

and this tends to zero by assumption.

q.e.d.

Theorem 2: *Let the above conditions be satisfied, in particular (7.1) - (7.3), let* f *be a bounded density, continuous at* x *and* $\mu(\{x\}) = 0$, *further we assume* $\mathrm{Rest}_H D \equiv 1$. *Define* $a_n$ *by*

$$a_n := \sup_{g \in G} \int_H \frac{T_n(gh)}{\chi(gh)} \, d\kappa(h)$$

*and let*

$$\lim_{n\to\infty} n/a_n^2 = \infty \quad \text{be satisfied.} \tag{9}$$

*Then* $(\hat{f}_n)$ *with kernels of the form* (8) *is locally consistent at* x.

Remark: *If* $\frac{T_n}{\chi}$ *is uniformly bounded by constants* $c_n$ *fulfilling* $n/c_n^2 \to \infty$, *then* (9) *is valid; if* G *is unimodular, this condition is implied by* $n/\|T_n\|_\infty^2 \to \infty$.

*Proof.*

$$s_n(x) = \sup_{y\in X} K_n(x,y) = \sup_{y \in X} \int_H \frac{T_n(g^{-1}g_1h_1)}{\chi(g_1h_1)} \, d\kappa(h_1) =$$

$$= \frac{1}{\chi(g)} \sup_{g_1 \in G} \int_H \frac{T_n(g^{-1}g_1h_1)}{\chi(g^{-1}g_1h_1)} \, d\kappa(h_1) = \frac{1}{\chi(g)} \cdot a_n .$$

This equation is independent of the representative of x, since, if $g = g'h$ with $h \in H$, we have

$$\sup_{y \in X} \int_H \frac{T_n((g'h)^{-1}g_1h_1)}{\chi(g_1h_1)} \, d\kappa(h_1) =$$

$$= \frac{1}{\chi(g'h)} \cdot \sup_{g_1 \in G} \int_H \frac{T_n(h^{-1}(g')^{-1}g_1h_1)}{\chi(h^{-1}(g')^{-1}g_1h_1)} \, d\kappa(h_1) =$$

$$= \frac{1}{\chi(g')\cdot\chi(h)} \cdot a_n = \frac{1}{\chi(g')} \cdot a_n, \text{ since } \chi(h) = 1/D(h) = 1.$$

The measure $\mu$ derived by Haar measure being regular, Theorem 1 applies.

q.e.d.

Corollary 3: *Under the conditions of Theorem 2, let* f *be bounded and continuous at* x*, moreover* $\sum_{n=1}^{\infty} \exp(-\delta n/a_n^2) < \infty$ *for every* $\delta > 0$. *Then* $(\hat{f}_n)$ *is locally strongly consistent at* x.

## References

[ 1] BERAN, R.J.: Testing for uniformity on compact homogeneous space. J. Appl. Probab. 5 (1968), 177-195.

[ 2] GAAL, St.A.: Linear Analysis and Representation Theory. Springer, Berlin, 1973.

[ 3] HOEFFDING, W.: Probability inequalities for sums of bounded random variables. J. Amer. Statist. Assoc. 58 (1963), 13-30.

[ 4] MARDIA, K.V.; R.J.GADSEN: A small circle of best fit for sherical data and areas of vulcanism. Appl. Statist. 26 (1977), 238-245.

[ 5] NADARAYA, Ė.A.: On non-parametric estimates of density functions and regression curves. Teor. Verojatnost. i Primenen. 10 (1965), 199-203.

[ 6] REITER, H.: Classical Harmonic Analysis and Locally Compact Groups. Oxford Mathematical Monographs, Oxford, 1968.

[ 7] ROSENBLATT, M.: Remarks on some nonparametric estimates of a density function. Ann. Math. Statist. 27 (1956), 832-837.

[ 8] RUKHIN, A.L.: On the estimation of a rotation parameter on the sphere. Zap. Naučn. Sem. Leningrad, Otdel. Mat. Inst. Steklov (LOMI), 41 (1974), 94-104.

[ 9] SCHNEIDER, B.; W.WERTZ: Statistical density estimation - a bibliography. Internat. Statist. Rev. 47 (1979), 155-175.

[10] SCHUSTER, E.F.: Estimation of a probability density function and its derivatives. Ann. Math. Statist. 40 (1969), 1187-1195.

[11] SCHUSTER, E.F.: Note on the uniform convergence of density estimates. Ann. Math. Statist. 41 (1970), 1347-1348.

[12] WATSON, G.S.: Orientation statistics in geology. Bull. Geol. Inst. Univ. Upsala N.S. 2:9 (1970), 73-89.

[13] WERTZ, W.: Über ein nichtparametrisches Schätzproblem. Metrika 26 (1979), 157-167.

# NON-ERGODIC STATIONARY INFORMATION SOURCES

Karel Winkelbauer

Charles University, Prague

Throughout the sequel $(X,F,m)$ means a probability space and $T$ is an invertible measure-preserving transformation of the space, which will be referred to as its automorphism. The additive group of integers is denoted by $I$; hence $(T^i, i \in I)$ means the cyclic group of automorphisms associated with $T$. For a class of sets $C \subset F$ the notation $\sigma C$ means the sub-$\sigma$-algebra of $F$ that is generated by $C$.

To define the basic concept of this paper, a modified version of topological entropy, we shall restrict ourselves to the class $Z$ of finite measurable partitions of the space under consideration, which constitutes a lattice with respect to the relation $P \leq Q$ (partition $Q$ is a refinement of partition $P$); in what follows we shall set

$$P_T^n = \bigvee_{i=0}^{n-1} T^{-i}P \quad \text{for} \quad P \in Z.$$

Given $0 < \varepsilon < 1$, we define

$$L(\varepsilon, P) = \min \left\{ \mathrm{card}(C) : C \subset P, \sum_{E \in C} m(E) > 1 - \varepsilon \right\},$$

$$H_T(P) = \sup_{0<\varepsilon<1} \limsup_n (1/n) \log L(\varepsilon, P_T^n),$$

$$H(T) = \sup_{P \in Z} H_T(P).$$

We shall call $H(T)$ the asymptotic rate of automorphism $T$; it is shown in [2] that the asymptotic rate $H(T)$ is a numerical invariant of the dynamical system $(X,F,m,T)$. If $h(T)$ denotes the entropy of automorphism $T$ then $H(T) \geq h(T)$; especially,

$$H(T) = h(T) \quad \text{if } T \text{ is ergodic.}$$

We may refer to $H_T(P)$ as the asymptotic rate of partition $P$ with respect to $T$. As easily seen, the definition of $H_T(P)$ makes sense for any countable measurable partition $P$; moreover,

$$H_T(P) \leq H(T) \quad \text{if} \quad h(P) < +\infty$$

(here $h(P)$ is the entropy of partition $P$). A countable measurable partition $G$ is called a generator for $T$ if

$$\sigma\{T^i E: \ E \in G, \ i \in I\} = F \mod 0;$$

if $G$ is a generator for $T$ then

$$H_T(G) = H(T) \quad \text{supposed that } h(G) < +\infty \ .$$

The properties of the asymptotic rate summarized above are all proved in [2] ; there it is also shown that the numerical value of the asymptotic rate depends on the non-atomic part of the measure considered only.

NOTE. The asymptotic rate was studied by the author first in 1962 for the case of two-sided shifts, i.e. for stationary discrete information sources; cf. the quotations in [2] .

A countable set $A$ of cardinality $>1$ being given, let $S = S_A$ be the shift in $A^I$ (defined by $(Sx)_i = x_{i+1}$), and let $m$ be a shift-invariant probability measure defined on the $\sigma$-algebra $F_A$ generated by the class of finite-dimensional cylinders. Then the dynamical system $(A^I,F_A,m,S_A)$ observed from the viewpoint of information theory is called a stationary (discrete) information source and usually denoted as $[A,m]$ ; the asymptotic rate of the shift will be referred to in what follows as the asymptotic rate of the stationary information source under consideration. The decisive role which plays the numerical value of the asymptotic rate in transmitting the information produced by a stationary (in general, non-ergodic) information source, may be recognized from the following basic theorem.

THEOREM. If $[A,m]$ is a stationary information source such that $m$ is a non-atomic measure, and if $H$ is the asymptotic rate of the source having the property that $H < +\infty$ , then there is a stationary information source $[A_o,m_o]$ which is isomorphic with the source $[A,m]$ and such that

$$\exp H < \operatorname{card}(A_o) \leqslant \exp H + 1.$$

Moreover, a stationary source $[A_o,m_o]$ such that $\operatorname{card}(A_o) < \exp H$ cannot be isomorphic with $[A,m]$ .

Under the isomorphism between stationary sources $[A,m]$ and $[A_o,m_o]$ it is understood that the shifts $S=S_A$ and $S_o=S_{A_o}$ are isomorphic in the usual sense, i.e. that there is a mod 0 isomorphism $t$ between $A^I$ and $A_o^I$ with the property that $S_o t = tS$ and $m_o = mt^{-1}$. In the proof of the preceding theorem (to appear in [3] ) there is given an explicit construction of such an isomorphism $t$ .

The theorem stated above yields a basis for the proof of a general theorem on dynamical systems which are aperiodic. Recall that an automorphism $T$ of a probability space $(X,F,m)$ is periodic for $E \in F$ if there is a positive integer $p$ such that $T^pF = F$ mod 0 for any $F \subset E$; $T$ is said aperiodic if it is periodic only for sets of measure zero.

THEOREM. If $T$ is an aperiodic automorphism of a countably generated probability space such that its asymptotic rate $H(T)$ is finite, then there is a finite generator for $T$ of cardinality not exceeding $\exp H(T) + 1$.

Conversely, any finite generator for $T$ is of cardinality not less than $\exp H(T)$; moreover,

$$H(T) = \inf\{h(G): \ G \text{ is a generator for } T\},$$

where h(G) is the entropy of partition G.

In case that T is ergodic (where H(T) equals the entropy of T, as seen above) the statement of the theorem coincides with Krieger s theorem proved in [1] . The author s proof of the general theorem just stated is based on quite a different method than that of Krieger which is necessary because of the non-ergodicity of the automorphism (cf. [3] ). What is possible to obtain by making use of the method applied in [1] for the non-ergodic case, was shown by the author in [2] where it was established that the finiteness of the asymptotic rate guarantees the existence of finite generators.

## REFERENCES

[1] W. Krieger: On entropy and generators of measure preserving transformations. Trans.Amer.Math.Soc.149(1970),453-464.

[2] K. Winkelbauer: On the existence of finite generators for invertible measure-preserving transformations. Comm.Math.Univers. Carolinae 18(1977),789-812.

[3] K. Winkelbauer: Finite generators of minimum cardinality for invertible measure-preserving transformations. To appear in:Comm. Math.Univers.Carolinae 21(1980).

## AUTHOR'S ADDRESSES

Jiři ANDĚL
Matematický Ústav
University Karlovy

Sokolovská 83
CS-18600 Praha 8

---

Dragan BANJEVIĆ
Odsek za Matematiku
Prirodno-matematicki Fakultet

Studentski trg 16
YU-11000 Beograd

---

Nader Labib BASSILY

Donáti u. 67.III.13
H-1015 Budapest

---

Boguslawa BEDNAREK-KOZEK
Uniwersytet Wrocławski
Instytut Matematiyczny

pl. Grundwaldski 2/4
PL-50-348 Wrocław

---

Tomasz BYCZKOWSKI
Instytut Matematyki
Politechniki Wrocławskiej

Wybreże St, Wyspiánskiego 27
PL-50-370 Wrocław

---

Endre CSÁKI
Magyar Tudományos Akadémia
Matematikai Kutató Intézete

Reáltanoda utca 13-15
H-1053 Budapest

---

Antónia FÖLDES
Magyar Tudományos Akadémia
Matematikai Kutató Intézete

Reáltanoda utca 13-15
H-1053 Budapest

---

Peter GERL
Institut für Mathematik
Universität Salzburg

Petersbrunnerstr. 19
5020 Salzburg

Erhard GLÖTZL
Institut für Statistik
Johannes-Kepler-Universität

Altenberger Straße 69
A-4040 Linz

---

Wilfried GROSSMANN
Institut für Statistik
und Informatik

Rathausstraße 19/3
A-1010 Wien

---

Béla GYIRES
Kossuth Lajos Tudományegetem
Matematikai Intézete

pf. 12
H-4010 Debrecen

---

T. INGLOT
Instytut Matematyki
Politechniki Wrocławskiej

Wybreże St. Wyspiánskiego 27
PL-50-370 Wrocław

---

A. IVANYI
Eötvös Loránd Tudományegyetem
Numerikus és Gépi Matematikai
Tanszék

Múzeum krt. 6-8
H-1088 Budapest

---

Zoran IVKOVIĆ
Odsek za Matematiku
Prirodno-matematicki Fakultet

Studentski trg 16
YU-11000 Beograd

---

Imre KÁTAI
Eötvös Loránd Tudományegetem
Numerikus és Gépi Matematikai
Tanszék

Múzeum krt. 6-8
H-1088 Budapest

---

Berthold RAUCHENSCHWANDTNER
Institut für Statistik
Johannes-Kepler-Universität

Altenberger Straße 69
A-4045 Linz

Franz KONECNY
Institut für Mathematik
Universität f. Bodenkultur

Gregor Mendel-Str. 33
A-1180 Wien

---

Andrzej KOZEK
Instytut Matematyczny
Polskiej Akademii Nauk

Kopernika 18
PL-51-617 Wrocław

---

Norbert KUSOLITSCH
Inst.f.Statistik u.Wahrscheinlichkeitstheorie, TU Wien

Argentinierstr. 8/7
A-1040 Wien

---

József MOGYORÓDI
Budapesti Eötvös Loránd
Tudómanyegyetem
Valószinüségszámitási Tanszéke

Múzeum krt. 6-8
H-1088 Budapest

---

Tamás MÓRI
Budapesti Eötvös Loránd
Tudományegyetem
Valószinüségszámitási Tanszéke

Múzeum krt. 6-8
H-1088 Budapest

---

Tibor NEMETZ
Magyar Tudományos Akadémia
Matematikai Kutató Intézete

Reáltanoda utca 13-15
H-1053 Budapest

---

Georg PFLUG
Institut für Statistik
und Informatik

Rathausstr. 19/3
A-1010 Wien

---

Lidia REJTŐ
Magyar Tudományos Akadémia
Matematikai Kutató Intézete

Reáltanoda utca 13-15
H-1053 Budapest

---

Pál RÉVÉSZ
Magyar Tudományos Akadémia
Matematikai Kutató Intézete

Reáltanoda utca 13-15
H-1053 Budapest

---

Tomasz ROLSKI
Uniwersytet Wrocławski
Instytut Matematyczny

pl. Grundwaldski 2/4
PL-50-384 Wrocław

---

Leszek RUTKOWSKI
ul. Zielinskiego 26/16

PL-53-534 Wrocław

---

Károly SARKADI
Magyar Tudományos Akadémia
Matematikai Kutató Intézete

Reáltanoda utca 13-15
H-1053 Budapest

---

Ferenc SCHIPP
Eötvös Loránd Tudományegetem
Numerikus és Gépi Matematikai
Tanszék

Múzeum krt. 6-8
H-1088 Budapest

---

Gábor J. SZÉKELY
Budapesti Eötvös Loránd
Tudományegyetem
Valószinüségszámitási Tanszéke

Múzeum krt. 6-8
H-1088 Budapest

---

Dominik SZYNAL
Uniwersytet M.C.-S.
Instytut Matematyki

ul. Nowotki 10
PL-20-031 Lublin

---

Jószef TOMKÓ
Kossuth Lajos Tudományegetem
Matematikai Intézete

pf. 12
H-4010 Debrecen

---

István VINCZE
Magyar Tudományos Akadémia
Matematikai Kutató Intézete

Reáltanoda utca 13-15
H-1053 Budapest

---

Wolfgang WERTZ
Inst. f. Statistik u. Wahrscheinlichkeitstheorie, TU Wien

Argentinierstr. 8/7
A-1040 Wien

---

Karel WINKELBAUER
Matematický Ústav
University Karlovy

Sokolovská 83
CS-18600 Praha 8

# Lecture Notes in Statistics

Vol. 1: R. A. Fisher: An Appreciation. Edited by S. E. Fienberg and D. V. Hinkley. xi, 208 pages, 1980.

Vol. 2: Mathematical Statistics and Probability Theory. Proceedings 1978. Edited by W. Klonecki, A. Kozek, and J. Rosiński. xxiv, 373 pages, 1980.

Vol. 3: B. D. Spencer, Benefit-Cost Analysis of Data Used to Allocate Funds. viii, 296 pages, 1980.

Vol. 4: E. A. van Doorn: Stochastic Monotonicity and Queueing Applications of Birth-Death Processes. vi, 118 pages, 1981.

Vol. 5: T. Rolski, Stationary Random Processes Associated with Point Processes. vi, 139 pages, 1981.

Vol. 6: S. S. Gupta and D.-Y. Huang, Multiple Statistical Decision Theory: Recent Developments. viii, 104 pages, 1981.

Vol. 7: M. Akahira and K. Takeuchi, Asymptotic Efficiency of Statistical Estimators. viii, 242 pages, 1981.

Vol. 8: The First Pannonian Symposium on Mathematical Statistics. Edited by P. Révész, L. Schmetterer, and V. M. Zolotarev. vi, 308 pages, 1981.

# Springer Series in Statistics

L. A. Goodman and W. H. Kruskal, Measures of Association for Cross Classifications. x, 146 pages, 1979.

J. O. Berger, Statistical Decision Theory: Foundations, Concepts, and Methods. xiv, 420 pages, 1980.

R. G. Miller, Jr., Simultaneous Statistical Inference, 2nd edition. 300 pages, 1981.

P. Brémaud, Point Processes and Queues: Martingale Dynamics. 352 pages, 1981.

# Lecture Notes in Mathematics

Selected volumes of interest to statisticians and probabilists:

Vol. 532: Théorie Ergodique. Actes des Journées Ergodiques, Rennes 1973/1974. Edited by J.-P. Conze, M. S. Keane. 227 pages, 1976.

Vol. 539: Ecole d'Ete de Probabilités de Saint-Flour V–1975. A. Badrikian, J. F. C. Kingman, J. Kuelbs. Edited by P.-L. Hennequin. 314 pages, 1976.

Vol. 550: Proceedings of the Third Japan-USSR Symposium on Probability Theory. Edited by G. Maruyama, J. V. Prokhorov. 722 pages, 1976.

Vol. 566: Empirical Distributions and Processes. Selected Papers from a Meeting at Oberwolfach, March 28–April 3, 1976. Edited by P. Gänssler, P. Revesz. 146 pages, 1976.

Vol. 581: Séminaire de Probabilités XI. Université de Strasbourg. Edited by C. Dellacherie, P. A. Meyer, and M. Weil. 573 pages, 1977.

Vol. 595: W. Hazod, Stetige Faltungshalbgruppen von Warscheinlichkeitsmassen und erzeugende Distributionen. 157 pages, 1977.

Vol. 598: Ecole d'Ete de Probabilités de Saint-Flour VI–1976. J. Hoffmann-Jorgenser T. M. Ligett, J. Neveu. Edited by P.-L. Hennequin. 447 pages, 1977.

Vol. 636: Journées de Statistique des Processus Stochastiques, Grenoble 1977. Editec by Didier Dacunha-Castelle, Bernard Van Cutsem. 202 pages, 1978.

Vol. 649: Séminaire de Probabilités XII. Strasboug 1976–1977. Edited by C. Dellacherie, P. A. Meyer, and M. Weil. 805 pages, 1978.

Vol. 656: Probability Theory on Vector Spaces, Proceedings 1977. Edited by A. Weron. 274 pages. 1978.

Vol. 672: R. L. Taylor, Stochastic Convergence of Weighted Sums of Random Elements in Linear Spaces. 216 pages, 1978.

Vol. 675: J. Galambos, S. Kotz, Characterizations of Probability Distributions. 169 pages, 1978.

Vol. 678: Ecole d'Ete de Probabilités de Saint-Flour VII–1977. D. Dacunha-Castelle, H. Heyer, and B. Roynette. Edited by P.-L. Hennequin. 379 pages, 1978.

Vol. 690: W. J. J. Rey, Robust Statistical Methods. 128 pages, 1978.

Vol. 706: Probability Measures on Groups, Proceedings 1978. Edited by H. Heyer. 348 pages, 1979.

Vol. 714: J. Jacod, Calcul Stochastique et Problemes de Martingales. 539 pages, 1979

Vol. 721: Seminairé de Probabilités XIII. Proceedings, Strasbourg, 1977/78. Edited b C. Dellacherie, P. A. Meyer, and M. Weil. 647 pages, 1979.

Vol. 794: Measure Theory, Oberwolfach 1979, Proceedings, 1979. Edited by D. Kolzöw. 573 pages, 1980.

Vol. 796: C. Constantinescu, Duality in Measure Theory. 197 pages, 1980.

Vol. 821: Statistique non Paramétrique Asymptotique, Proceedings 1979. Edited by J.-P. Raoult. 175 pages, 1980.

Vol. 828: Probability Theory on Vector Spaces II, Proceedings 1979. Edited by A. Weron. 324 pages, 1980.